Operantive Remediology and Gymnopraxis for Premenarchic Nosological Signs in Hebetics

Lyric Dexter

Operantive Remediology and Gymnopraxis for Premenarchic Nosological Signs in Hebetics

First Edition November 2025

Written by Lyric Dexter

TABLE OF CONTENTS

Sr. No.	Content	Page No.
1.	**INTRODUCTION AND OBJECTIVES (Chapter-I)**	**1-4**
2.	**REVIEW OF LITERATURE (Chapter-II)**	**5-34**
2.1.	Diagnostic guidelines of premenstrual syndrome	5-8
2.2.	Global burden of premenstrual syndrome	8-10
2.3.	Indian burden of premenstrual syndrome	10-13
2.4.	Factors influencing the premenstrual syndrome	13
2.4.1.	Stress and anxiety in adolescents with premenstrual syndrome	13
2.4.2.	Lifestyle factors	14
2.4.3.	Dietary influence	14
2.4.4.	Sleep patterns	14-15
2.4.5	Mood disturbances	15
2.4.6.	Exercise adherence in adolescents	15
2.5.	Symptoms of premenstrual syndrome	16
2.5.1.	Physical symptoms	16
2.5.2.	Emotional symptoms	17
2.5.3.	Behavioral symptoms	17
2.5.4	Comorbidities and overlapping conditions	18
2.6.	Interventions for premenstrual syndrome	18
2.6.1.	Pharmacological interventions for premenstrual syndrome	18-19
2.6.2.	Hormonal therapies	19-20
2.6.3.	Aerobic exercise	20-21
2.6.4.	Swimming	21-22

2.6.5.	Mindfulness	22
2.6.6.	Life style change	22-23
2.6.7.	Cognitive behavioral therapy	23-24
2.6.8.	Group cognitive behavioral therapy	24
2.6.9.	Multidisciplinary approach	24-25
2.6.10.	Yoga	25-26
2.6.11.	Aroma therapy	26
2.6.12.	Acupuncture	26
2.6.13.	Dietary supplements.	27
2.6.14.	Homeopathy	27
2.6.14.	Endocrinological influence on premenstrual syndrome	28
2.6.14.1.	Estrogen and progesterone fluctuations	28
2.6.14.2	Role of Prolactin and cortisol	28-29
2.5.14.3.	Thyroid dysfunction	29
2.6.15.	Cultural perspectives	29-30
2.6.16.	Education	30
2.6.17	Peer support	30-31
2.6.18.	Comparative efficacy of cognitive behavioral therapy and aerobic exercise in premenstrual syndrome management	31-32
2.7.	Barriers to treatment	32
2.7.1.	Awareness and education gaps	32
2.7.2.	Cultural stigma and social norms	32-33
2.7.3.	Financial constraints	33
2.7.4.	Limited access to healthcare providers	33
2.7.5.	Addressing barriers to treatment	33-34
3.	**MATERIAL AND METHODS (Chapter –III)**	**35-65**
3.1.	Study design	35

3.2.	Study population	35
3.3.	Pilot study	35
3.4.	Sample size	35-36
3.5.	Inclusion criteria	36
3.6.	Exclusion criteria	36
3.7.	Selection protocol	36-37
3.8.	Material used	37
3.8.1	Weighing machine	37
3.8.2.	Stature meter	37
3.8.3.	Sociodemographic data sheet	38
3.8.4.	Daily record of severity of symptoms problem diary	38
3.8.5.	Premenstrual syndrome screening tool For adolescent	38-39
3.8.6.	Premenstrual syndrome scale	39
3.8.7.	Pain disability index	39-40
3.8.8.	Health related short form-36 questionnaire (RAND)	40
3.8.9.	Beck depression inventory-II	40
3.8.10.	Beck anxiety inventory	40-41
3.9.	Diagnostic assessment	41
3.10.	Demographic variables	41-42
3.10.1.	Primary outcomes	42
3.10.2.	Secondary outcomes	42
3.11.	Enrolment procedure	42-44
3.12.	Intervention protocols	44
3.12.1.	Cognitive behavioral therapy protocol	44-49
3.12.2	Aerobic exercise protocol	50
	a. Warm up phase	50-52

	b. Aerobic exercises	53 – 56
	c. Cool down phase	57
3.13.	Follow up	57 - 58
3.14.	Statistical analysis	59
4.	**RESULTS (Chapter –IV)**	**60 – 65**
4.1.	Sociodemographic characteristics of study participants	60– 61
4.1.1.	Prevalence of premenstrual syndrome	62 – 63
4.1.2.	Age of attainment of menarche	64
4.1.3.	Body mass index of the participants	64
4.1.4.	Span of menstruation cycle	64
4.1.5.	Perceived menstruation flow	65
4.1.6.	Habit of changing absorbent	65
4.1.7.	Education status of the parents of the participants	65
4.1.8.	Employment status of the parents of the participants	65
4.2.	Outcome measures for CBT and aerobic exercises	69
4.2.1.	Pre-intervention scores	69 - 72
4.2.2.	Post-interventional scores	72
4.2.3.	Comparative effect of CBT and aerobic exercise	78-80
4.2.4.	Effect size of CBT and aerobic exercise interventions	84-89
5.	**DISCUSSION (Chapter-V)**	**90 – 104**
6.	**SUMMARY AND CONCLUSION (Chapter-VI)**	**105-107**

LIST OF FIGURES

Figure No.	Description of figure	Page No.
4.1.3.	Distribution of participants according to age of attainment of menarche	**66**
4.1.4.	Distribution of participants according to body mass index	**66**
4.1.5.	Distribution of participants according to span of menstruation cycle	**67**
4.1.5.	Distribution of participants according to perceived menstruation flow	**67**
4.1.7.	Distribution of participants according to habit of changing absorbent	**68**
4.1.8.	Distribution of the participants according to the education status of parents	**68**
4.1.9.	Distribution of the participants according to employment of their parents	**73**
4.2.1.	Pre-intervention scores of pain disability index, beck anxiety inventory and beck depression inventory-II of group cognitive behavioral therapy and group aerobic exercise.	**73**
4.2.2.	Pre-intervention scores for short form 36 (Sf36) domains of cognitive behavioral therapy and aerobic exercise group	**74**
4.2.2.1	Pre and post-intervention scores of pain disability index, beck anxiety inventory and beck depression inventory-II of cognitive behavioral therapy (CBT) group	**74**
4.2.2.2.	Pre-intervention and post-intervention scores of domains of Short form 36 (Sf36) for participants of cognitive behavioral therapy (CBT) group	**78**
4.2.2.3.	Pre-intervention and post-intervention scores of pain disability index, beck anxiety inventory and beck depression inventory-II of participants of aerobic exercise group	**78**
4.2 2.4.	Pre-intervention and post-intervention scores of domains of Short form 36 (Sf36) for aerobic exercise group	**79**
4.2 3.1.	Between the group comparison of Post-intervention scores of pain disability index, beck anxiety inventory and beck depression inventory-II of CBT and aerobic exercise groups.	**79**
4.2.3.2	Between the group comparisons for post-intervention scores of nine domains of Sf (36) of CBT and aerobic exercise groups	**80**

LIST OF TABLES

Table No.	Table description	Page No.
3.12.	Cognitive behavioral therapy protocol	**47-49**
3.13.	Warm up phase of aerobic exercises	**47-49**
3.14.	Details of aerobic exercise protocol	**50-56**
3.5.	Flow chart of methodology	**58**
4.1.	Details of methodology used for recruiting of participants for study	**61**
4.1.1.	List of common symptoms reported by the participants	**62**
4.1.2.	List of symptom's according to impact of severity reported by the participants	**63**
4.2.1.	Pre-intervention scores of pain disability index, beck anxiety inventory and beck depression inventory-II of group cognitive behavioral therapy and group aerobic exercise.	**69 – 70**
4.2.2.	Pre-intervention scores for short form 36 (Sf36) domains of cognitive behavioral therapy and aerobic exercise groups	**71– 72**
4.2.2.1.	Pre and post-intervention scores of pain disability index, beck anxiety inventory and beck depression inventory-II of cognitive behavioral therapy (CBT) group	**75**
4.2.2.2.	Pre-intervention and post-intervention scores for Short form 36 (Sf 36) domains of cognitive behavioral therapy (CBT) group	**76**
4.2.2.3.	Pre-intervention and post-intervention scores of pain disability index, beck anxiety inventory and beck depression inventory-II participants in aerobic exercise group	**77**
4.2.2.4.	Pre-intervention and post-intervention scores of domains of Short form 36 (Sf 36) for aerobic exercise group	**78**
4.2.3.1.	Between group comparisons of post- intervention mean scores of pain disability index, beck anxiety inventory and beck depression inventory-II of CBT and aerobic exercise groups.	**79**
4.2.3.2.	Comparisons of post-intervention scores of nine domains of Sf (36) after CBT and aerobic exercise groups	**80**

4.2.[illegible].1.	Effect Size of CBT intervention on pain disability index, beck anxiety inventory and beck depression inventory-II.	**84**
4.2.[illegible].2.	Effect Size of aerobic exercise intervention scores on pain disability index, beck anxiety inventory and beck depression inventory-II	**85**
4.2.[illegible].3.	Effect size of CBT intervention on nine domains of Sf (36)	**86**
4.2.[illegible].4.	Effect Size of aerobic exercise intervention on Sf (36) domains	**87 - 88**
4.2.[illegible].5.	Comparison of effect size for CBT and aerobic exercise groups	**88- 89**

LIST OF APPENDICES

Sr. no.	Description of appendix	Appendices No.
1	Sociodemographic data sheet	A
2	Daily record of severity of symptoms problem diary	B
3	Premenstrual syndrome screening tool for adolescents	C
4	Premenstrual syndrome scale	D
5	Pain disability index	E
6	Short form 36 Sf (36)	F
7	Beck depression inventory-II	G
8	Beck anxiety inventory	H
9	Consent	I

LIST OF ABBREVIATION

Sr.No.	Full form	Abbreviation
1.	Premenstrual syndrome	PMS
2.	Premenstrual dysphoric disorder	PMDD
3.	Diagnostic and statistical manual of mental disorders fifth edition	DSM-V
4.	American psychiatric association	APA
5.	American congress of obstetricians and gynecologists	ACOG
6.	Non-steroidal anti-inflammatory drugs	NSAIDs
7.	Cognitive behavioral therapy	CBT
8.	World health organization	WHO
9.	Premenstrual tension syndrome	PMTS
10.	International classifications of disabilities-10	ICD-10
11.	International diagnostic Interview	CIDI
12.	Daily record of severity of symptoms problem	DRSP
13.	Premenstrual symptoms screening tool for adolescents	PSST–A
14.	Premenstrual syndrome scale	PMSS
15.	Identification	ID
16.	Kilogram	Kg
17.	Centimeter	Cm
18.	Pain disability index	PDI
19.	Beck anxiety inventory	BAI
20.	Beck depression inventory -II	BDI-II
21.	Health-related quality of life	HRQoL
22.	Health-Related Short Form-36 Questionnaire	Sf (36)
23.	Body mass index	BMI
24.	Gamma-aminobutyric acid	GABA
25.	Serotonin Reuptake Inhibitors	SSRIs

Introduction

The term "premenstrual tension" was first introduced by R. T. Frank in 1931, and later replaced with "premenstrual syndrome" (PMS) by Greene and Dalton in 1953 (Connolly, 2001). PMS is a common and complex condition that affects many females during their reproductive years, with particularly high rates reported among adolescents. It is defined as a cyclic pattern of emotional, physical, and behavioral symptoms that occur during the luteal phase of the menstrual cycle and subside with the onset of menstruation (World Health Organization, 1992). Although its precise cause is still unknown, PMS is widely believed to arise from a combination of hormonal fluctuations, neurotransmitter changes, genetic factors, and environmental influences. Adolescents are especially vulnerable due to the hormonal changes associated with puberty. Reported prevalence rates range from 30% to 80%, with variations influenced by lifestyle, cultural background, and socioeconomic conditions (Arafa et al., 2018; Acharya et al., 2006). Globally, the prevalence of PMS has been reported between 5% and 90%, while rates among adolescents range from 0% to 53%, often affecting their quality of life (Dambhare et al., 2012; Lee et al., 2006). Some studies even report 100% prevalence among young and middle-aged women in western India (Beahmbatt et al., 2013).

Lifestyle patterns among adolescents have also been linked to menstrual problems (Nagi et al., 2019). Numerous studies confirm that PMS interferes with daily activities and is a major cause of absenteeism from work or school (Shamnani et al., 2018; Dennerstein et al., 2010; Taghizadeh et al., 2008). PMS includes a wide range of symptoms—more than 150 have been documented—such as abdominal bloating, headaches, breast tenderness, irritability, mood swings, anxiety, and depression, but no single diagnostic test exists (Derman et al., 2004). These symptoms can significantly disrupt daily life, affecting academic performance, social interaction, and overall well-being (Antai et al., 2004; Ghanbari et al., 2008). This disruption is especially concerning for adolescents, who are already undergoing major physical and emotional changes. Research shows that PMS is a leading contributor to school absenteeism, reduced academic productivity, and strained relationships (Asadi et al., 2020).

social interactions, and overall quality of life in affected individuals Antai et al., (2004) and Ghanbari et al., (2008). This disruption is concern particularly in adolescents, a group already navigating significant physical and emotional changes. Research highlights that PMS symptoms are a major cause of absenteeism from school, decreased academic productivity, and strained interpersonal relationships Asadi et al., (2020).

Premenstrual syndrome in adolescents is further influenced by lifestyle factors such as poor dietary habits, physical inactivity, and elevated stress levels. Studies suggest that inadequate nutrition exacerbates premenstrual syndrome symptoms, emphasizing the critical role of dietary patterns in menstrual health Ruprai et al., (2015) and Jami, 2016). Deficiencies in calcium, vitamin D, and magnesium have been linked to heightened symptom severity, while balanced nutrition has been shown to alleviate symptoms Bakhshani et al., (2009). Similarly, psychological factors, including difficulties in emotional regulation and stress management challenges, have been implicated in worsening PMS symptoms. Adolescents who habitually suppress emotions are reported to experience more severe symptoms than those who adopt positive coping strategies Lawal et al., (2019) and Wu et al., (2016).

In terms of management, pharmacological interventions such as hormonal therapies and non-steroidal anti-inflammatory drugs have traditionally been employed to provide symptomatic relief. However, these treatment modalities are often associated with side effects and have only limited long-term efficacy which make non-pharmacological interventions increasingly popular (Modzelewski et al., (2021). Aerobic exercise and CBT are two widely recognized interventions that have demonstrated efficacy in alleviating both physical and psychological aspects of PMS Lathavenkatesan & Annapoorani, (2019).

The benefits of aerobic exercise in PMS management are well-documented, in studies demonstrating significant reduction in symptoms such as cramps, fatigue, and mood swings Ravichandran, (2018) and Janakiraman, (2019). The physiological mechanisms underlying these improvements include increased endorphin release, enhanced blood circulation, and reduced stress hormone levels. Moreover, regular exercise promotes a sense of well-being and self-confidence, empowering adolescents to take control of their health. Therefore, exercise routines tailored to individual preferences and abilities are particularly effective in achieving sustained benefits.

Cognitive behavioral therapy is a structured psychological intervention that aims to address the cognitive distortions and maladaptive behaviors associated with PMS. By fostering self-awareness and emotional regulation, CBT equips adolescents with practical coping mechanisms to manage psychological symptoms such as anxiety, irritability, and depression. Studies have consistently shown that CBT not only reduces symptom severity but also enhances overall quality of life Abootalebi et al., (2017). For instance, group-based CBT interventions have been particularly effective in fostering peer support and shared learning among adolescents, amplifying the therapeutic impact Hunter et al., (2002) and Wise et al., (2019).

Despite the documented efficacy of aerobic exercise and CBT, few studies have directly compared their relative effectiveness or explored the potential for their combined intervention Rethorst & Trivedi (2013). Preliminary evidence suggests that these approaches may complement each other, with aerobic exercise primarily addressing physical symptoms and CBT focusing on psychological aspects. An integrative approach combining lifestyle modifications, such as regular aerobic exercise and dietary adjustments, with psychological interventions like CBT, could provide holistic benefits by addressing both the physical symptoms and emotional challenges of PMS, particularly in adolescents navigating hormonal and psycho-social changes. This is particularly relevant in backward regions of Punjab like Bathinda, where cultural norms and limited access to healthcare often exacerbate the challenges faced by young girls with PMS. With the changes and advancement of trends in PMS management, it is essential to reassess existing approaches and establish new evidence-based recommendations tailored to the adolescent population.

Despite the documented efficacy of both aerobic exercise and CBT in managing symptoms of PMS, limited research directly compares these interventions, particularly among adolescent populations. Moreover, there is a lack of studies focused on conservative treatment strategies for PMS in the context of Southern Punjab. Existing evidence supports the effectiveness of non-pharmacological interventions for managing PMS symptoms, highlighting the importance of approaches like aerobic exercise and CBT. Studies have shown that regular aerobic exercise helps alleviate both physical symptoms, such as bloating and fatigue, and emotional symptoms, including mood swings and irritability Zoghi et al. (2010). Similarly, CBT has been proven effective in addressing the psychological aspects of PMS, such as anxiety and depression, by helping individuals reframe negative thoughts and develop healthy coping mechanisms O'Neill et al. (2009). Additionally, dietary modifications, including increased intake of magnesium and omega-3 fatty acids, have been shown to reduce symptoms like irritability and fatigue Haleem et al., (2010), while relaxation techniques like yoga and mindfulness have been effective in alleviating both physical and emotional symptoms Manzoli et al., (2012). However, despite this, there is a notable gap in research comparing the direct effectiveness of aerobic exercise and CBT, especially in adolescent girls. By addressing these important gaps, the present study primarily seeks to evaluate and compare the effectiveness of aerobic exercise and CBT as treatment strategies for managing PMS in adolescent girls, contributing to the development of evidence-based interventions tailored to the need of adolescent girls and improving their quality of life with specific objectives as given below.

- Evaluating the effect of cognitive behavior therapy and aerobic exercises on pain disability index and quality of life of adolescents.
- Comparing the effectiveness of cognitive behavior therapy and aerobic exercises on premenstrual syndrome among adolescent girls.

2. REVIEW OF LITERATURE

Premenstrual syndrome is a multi-factorial condition characterized by emotional, physical, and behavioral symptoms occurring in the luteal phase of the menstrual cycle that subsides after onset of menstruation. Several studies have explored interventions to alleviate Premenstrual syndrome symptoms, yet the efficacy of different approaches continues to be a topic of interest. This study has been planned to compare the application of Cognitive Behavioral Therapy in conjunction with aerobic exercises to alleviate symptoms of premenstrual syndrome among adolescent girls. The available literature on the subject is reviewed below:

2.1. Diagnostic Guidelines of Premenstrual Syndrome

The World Health Organization (WHO) defines Premenstrual Tension Syndrome (PMTS) in the International Classifications of Disabilities-10 (ICD-10) as a condition predominantly defined by psychiatric symptoms. It requires only one mood or one physical symptom from specified lists and does not exclude symptoms resulting from the exacerbation of another disorder, nor does it require subjective dysfunction (World Health Organization, 1992). These definitions reflect the varying perspectives across different health organizations, emphasizing the complexity and diverse approaches to premenstrual syndrome diagnosis.

Wittchen et al., (2002) conducted a prospective longitudinal community survey of 1,488 young women aged 14–24, followed over 48 months. Utilizing the Composite International Diagnostic Interview (CIDI) and DSM-IV methods for diagnosing Premenstrual Syndrome and premenstrual dysphoric disorder (PMDD), the study revealed that while premenstrual symptoms were widespread among adolescents and young adults, PMDD was notably less prevalent. This study highlighted the high incidence of premenstrual syndrome in younger populations, but a lower prevalence of PMDD was observed because of its diagnosis requires more severe symptoms. A study by Yonkers et al., (2008), reported that a significant proportion of menstruating women has been affected by PMS, the PMDD is less common and requires intensive psychological intervention. Additionally, Kuehner, (2017) explored the role of hormonal and psychosocial factors effecting the severity of premenstrual syndrome and premenstrual dysphoric disorder. The premenstrual syndrome is influenced by hormonal disturbances, while psychological and emotional factors play a critical role in the premenstrual dysphoric disorder. Halbreich & Kuehner, **(2007)** suggested that in managing PMDD, a multi-disciplinary approach, including pharmacological and non-pharmacological interventions, is crucial. Similarly, Derman et al., (2004) used DSM-IV criteria in his study to evaluate the associated symptoms of PMS in adolescent girls and found that 71% of the participants experienced dysmenorrhea and premenstrual symptoms. This also shows that the co-occurrence of dysmenorrhea with PMS. This study further concluded that the disruptions to daily activities due dysmenorrhea in adolescents. Additionally, Harel, (2006) found that dysmenorrhea is one of the most common gynecological complaints among adolescent girls, further supporting the findings of Derman et al., (2004) regarding the frequent overlap of dysmenorrhea and premenstrual syndrome in this population. Omidvar et al., (2016) also noted that managing dysmenorrhea is critical in addressing the broader spectrum of premenstrual syndrome symptoms in adolescents.

The American Psychiatric Association (APA) defines premenstrual dysphoric disorder in the DSM-V as a severe form of premenstrual syndrome (PMS), characterized by the presence of at least five out of eleven symptoms, with at least one mood-related symptom during the luteal phase of the menstrual cycle. These symptoms must significantly disrupt daily functioning and cannot be attributed to the exacerbation of another disorder, such as depression or panic disorder (American Psychiatric Association, 2013).

The strict diagnostic criteria of APA for premenstrual dysphoric disorder, the American Congress of Obstetricians and Gynecologists (ACOG) adopted a broader and more inclusive approach to diagnosing PMS. American Congress of Obstetricians and Gynecologists defines premenstrual syndrome as the cyclical occurrence of a combination of mood and physical symptoms that manifest during the luteal phase of the menstrual cycle and resolve shortly after menstruation begins. Unlike the DSM-V criteria for PMDD, which emphasize the severity and exclusivity of mood symptoms, the American Congress of Obstetricians and Gynecologists criteria require the presence of at least one mood symptom (e.g., irritability, depression, or anxiety) and one physical symptom (e.g., bloating, breast tenderness, or headaches) from specified lists. This more inclusive diagnostic approach acknowledges the diversity in symptom presentation among individuals (American Congress of Obstetricians and Gynecologists, 2020).

According to the American College of Obstetrics and Gynecology and DSM-V guidelines, it is recommended that adolescent girls fill out a daily record of symptom severity over the previous two consecutive months. This approach helps in the accurate diagnosis and management of premenstrual syndrome by identifying patterns in symptom severity and tracking the presence of core symptoms. To meet the criteria for diagnosis, one core affective symptom (such as mood swings, irritability, or depression) and at least one core somatic symptom (like bloating, fatigue, or breast tenderness) must be present consistently over these two months. This method not only helps in identifying the severity and impact of symptoms on daily life but also in evaluating the effectiveness of interventions such as lifestyle changes, exercise, and psychological therapies like cognitive behavioral therapy. The American College of Obstetrics and Gynecology emphasizes the importance of using standardized tools like the daily record of symptom severity in clinical practice to ensure accurate diagnosis and effective management of premenstrual syndrome.

Importantly, the American Congress of Obstetricians and Gynecologists criteria do not exclude the possibility of symptom exacerbation due to underlying disorders such as depression, anxiety, or chronic pain conditions. Instead, the emphasis is placed on the cyclical nature of symptoms and the subjective experience of dysfunction. For a diagnosis to be made, symptoms must cause significant impairment in daily activities, such as work, school, or interpersonal relationships, as reported by the individual. Unlike PMDD, premenstrual syndrome as defined by American Congress of Obstetricians and Gynecologists does not require a minimum number of symptoms or formal psychiatric diagnostic tools. This approach allows for a more practical application in clinical settings, making it accessible for identifying premenstrual syndrome in a broader population of affected women.

By encompassing a wider range of symptoms and not strictly excluding comorbid conditions, the American Congress of Obstetricians and Gynecologists criteria recognize the complex interplay of hormonal, psychological, and environmental factors in the manifestation of premenstrual syndrome. This framework encourages personalized management strategies that address both the physical and emotional aspects of the condition.

2.2. Global Burden of Premenstrual Syndrome

The prevalence of premenstrual syndrome in adolescents varies across studies and regions, but it is generally reported that a significant proportion of adolescent girls experience some form of premenstrual syndrome symptoms. Studies indicate that the prevalence of premenstrual syndrome among adolescent girl's ranges from 40% to 90%, depending on factors such as age, socioeconomic status, and geographic location (Madan et al., 2021 and Sahoo et al., 2012). While the condition is widespread, the severity of symptoms varies greatly, with some girls experiencing mild discomfort and others facing debilitating symptoms that affect daily activities, including school performance and social interactions (Shamnani et al., 2018).

On a global scale, premenstrual syndrome imposes a substantial health burden, affecting the physical, emotional, and social well-being of women in reproductive age. Direkvand-Moghadam et al., (2014) conducted a systematic review to estimate the global prevalence of premenstrual syndrome, revealing significant variation ranging from 30% to 50%. The prevalence was notably higher in developing countries, where cultural stigmas surrounding menstruation often prevent women from seeking medical help. Study of Ranjbaran et al., (2017) found prevalence of PMS in Irani population was 80.4% among high school students, 68.9% among university students and 54.9% in general population. Alwafa et al., (2021) highlighted studies from the Middle East reporting premenstrual syndrome prevalence rates as high as 51.9% in Saudi Arabia. These variations were attributed to differences in socioeconomic status, education levels, and access to healthcare services.

Iacovides et al., (2015) explored the overlapping pathophysiology of premenstrual syndrome and dysmenorrhea, focusing on increased prostaglandin production and hormonal dysregulation. They reported that 64% of women experiencing dysmenorrhea also reported premenstrual symptoms. This comorbidity was associated with compounded physical symptoms like severe abdominal pain and fatigue, as well as heightened psychological symptoms such as irritability and anxiety. The authors noted that this dual burden contributed to a 23% increase in absenteeism among women in educational and professional settings.

Lifestyle interventions have shown promise in alleviating PMS symptoms. Farasati et al., (2015) conducted a randomized controlled trial involving 120 women aged 18–30, where participants engaging in regular aerobic exercise (three sessions per week for eight weeks) reported a 38% reduction in the severity of physical symptoms like bloating and breast tenderness and a 42% improvement in mood symptoms compared to the control group. This highlighted the potential of physical activity as a non-pharmacological approach to symptom management. A study conducted by Farideh Bastani and Sara Hashemi (2012) showed that web life style education can also significantly improve healthy nutrition, exercise/physical activity, behavioral and physical components of PMS symptoms Further, Yonkers & Simoni (2018) performed a meta-analysis of interventions for premenstrual syndrome, analyzing 15 studies with a total sample size of 1,200 participants. They found that combining CBT with regular exercise resulted in a 50% reduction in psychological symptoms such as irritability and anxiety, compared to a 20% reduction with exercise alone. Cognitive behavioral therapy was particularly effective in enhancing self-awareness and coping mechanisms, while aerobic activity alleviated physical symptoms such as fatigue and pain.

Bajalan et al., (2019) investigated dietary influences on premenstrual syndrome in a cross-sectional study of 300 women aged 20–35. Women with diets deficient in calcium and magnesium were found to be 2.8 times more likely to report severe premenstrual syndrome symptoms. Supplementation with calcium and magnesium for three months resulted in a 45% reduction in symptom severity, particularly in bloating and mood swings. The study emphasized the importance of micronutrient-rich diets in mitigating premenstrual syndrome-related distress.

Despite the high prevalence of premenstrual syndrome, particularly among adolescent girls, studies indicate that adolescent females experience some form of taboos around premenstrual syndrome Kumar et al., (2024), cultural taboos and lack of education about menstruation hinder timely intervention and management. Adolescents, especially in certain regions, may face challenges in seeking help due to stigma surrounding menstrual health. Addressing these barriers through education, improved healthcare access, and lifestyle modifications can significantly reduce the burden of premenstrual syndrome and enhance the quality of life for adolescent girls and women globally.

2.3. Indian Burden of Premenstrual Syndrome

Premenstrual syndrome represents a significant yet under-recognized health concern among adolescent girls in India. Its prevalence, influenced by cultural, socioeconomic, and lifestyle factors, highlights the need for targeted interventions. Despite affecting up to 60–80% of Indian adolescent girls, premenstrual syndrome is often under-reported due to cultural stigma and a lack of awareness. Premenstrual syndrome significantly affects the psychological well-being and social functioning of Indian adolescents.

Sahoo et al., (2012) conducted a cross-sectional study in urban population of India involving 800 adolescent girls aged 12–18. The study reported that 80% of participants experienced at least one PMS symptom, with 20% reporting symptoms severe enough to interfere with their daily functioning. Common symptoms included mood swings (68%), abdominal bloating (54%), and fatigue (62%). The study reported that 34% of participants experienced academic disruptions due to symptoms like mood swings and fatigue, while 22% reported strained relationships with peers and family members These findings emphasized the role of urban stress, poor dietary habits, and lack of physical activity in exacerbating premenstrual syndrome symptoms. Additionally, the study found a significant association between sedentary behavior and the severity of symptoms, highlighting the impact of lifestyle on premenstrual syndrome management.

A study by Garg et al., (2014) explored the correlation of symptoms among 148 Indian women and found a significant prevalence of premenstrual syndrome symptoms, such as mood swings, fatigue, and abdominal bloating, affecting daily life. This study highlighted the diverse symptomatology and the distress caused by premenstrual syndrome, particularly in terms of its impact on quality of life.

A comparative study conducted by Udaya et al., (2015) examined 1,200 adolescent girls in urban and rural settings. It revealed that while urban adolescents reported a higher prevalence of mild premenstrual syndrome symptoms (72% vs. 58%), rural adolescents experienced more severe symptoms (18% vs. 12%). Rural participants were more likely to suffer from inadequate nutrition, limited physical activity, and restricted healthcare access, contributing to worse symptomatology. For instance, only 10% of rural adolescents had knowledge of premenstrual syndrome management strategies compared to 35% in urban areas. Tarannum, et al., 2022 reported that PMS was 2.19 times more in late adolescence (15–19 years) and PMS had a statistically significant association with age. The study was a community-based cross-sectional study conducted in Secondary and Higher Secondary Schools (Girls) of Aligarh Muslim University (AMU), Aligarh. PMS was estimated as 37.7%, which was lower as compared to other studies from other part of India.

The numerous studies conducted by different authors revels variation in prevalence of PMS for instance, Singh et al., (2008) reported that 60.5% of the medical students suffered from PMS, Lakshmi et al., (2011) found the prevalence of PMS to be 67% whereas, Thakre et al., (2012) reported prevalence of PMS as 55.8%, Ramya et al. (2014) reported the prevalence of PMS in urban girls as 40.9% and in rural girls as 51.6% andSarkar et al., (2016) reported prevalence of PMS as 61.5%.

Cultural taboos surrounding menstruation in India pose significant challenges to premenstrual syndrome diagnosis and management. A qualitative study by Rani et al.,

(2017) involving focus group discussions with 150 adolescent girls from both urban and rural regions highlighted how menstruation is often viewed as a private matter, discouraging discussions about related symptoms. Participants reported feelings of shame and embarrassment, leading to delayed healthcare-seeking behavior. In rural areas, 72% of participants relied on traditional remedies or ignored their symptoms due to societal norms and lack of healthcare infrastructure. A study by Bhalerao et al., (2024) among 600 adolescent girls found that psychological stress, exacerbated by social pressures and stigma, increased the severity of premenstrual syndrome by 40%. In another study, Sundelin et al., (2019) examined the prevalence of premenstrual syndrome in a community-based sample of 600 Indian adolescent girls, revealing that 68% experienced symptoms such as mood swings, bloating, and fatigue, and 20% reported symptoms severe enough to interfere with daily activities. A study conducted by Deo et al., 2024 reported that anger outbursts, irritation and anxiety were revealed to be the most prevalent affective symptoms in adolescent girls and young women with PMS and PMDD. While in somatic symptoms, headache and weight gain were similarly significantly more common in persons with Moderate to severe PMS and PMDD compared to No/Mild PMS. BMI is also significantly associated with PMS and PMD, so it also affects weight management.

Madan et al., (2021) examined healthcare access and education about premenstrual syndrome among 1,000 adolescent girls in North India. The study found that only 24% of girls were aware of premenstrual syndrome as a medical condition, with the majority attributing symptoms to normal menstrual cycles or stress. Among those who sought treatment, 65% consulted family members or traditional healers, while only 12% accessed professional medical care. The authors noted that urban adolescents were more likely to receive education on menstrual health, whereas rural populations faced significant barriers due to limited outreach programs.

The Indian burden of premenstrual syndrome is influenced by complex interactions between cultural, socioeconomic, and lifestyle factors. Urban and rural differences highlight the need for tailored interventions to address unique challenges in each demographic. Increasing awareness, improving healthcare access, and addressing cultural taboos are critical to reducing the burden of premenstrual syndrome and improving the quality of life for adolescent girls in India.

2.4. Factors Influencing the Premenstrual Syndrome

The onset of premenstrual syndrome symptoms in adolescents is largely influenced by hormonal changes during puberty. Fluctuations in levels of estrogen and progesterone during the menstrual cycle are thought to play a central role in the onset and severity of premenstrual syndrome symptoms Yonkers et al., (2008). In adolescents, hormonal changes can be more erratic, potentially contributing to more severe symptoms of Premenstrual syndrome. Estrogen, which peaks in the follicular phase, and progesterone, which rises during the luteal phase, are the two key hormones implicated in premenstrual syndrome. An imbalance or sensitivity to these hormonal fluctuations may lead to symptoms such as mood swings, irritability, bloating, and abdominal cramps Steiner et al., (2003). Additionally, genetic factors may also contribute to susceptibility to premenstrual syndrome, although further research is needed to fully understand these mechanisms Pasquali et al., (2022) & Treloar et al., (2022)

2.4.1. Stress and Anxiety in Adolescents with Premenstrual Syndrome

Stress and anxiety have been identified as significant exacerbating factors for premenstrual syndrome symptoms in adolescent girls. During adolescence, girls undergo a variety of psychological and emotional changes, and when coupled with premenstrual syndrome, stress can significantly heighten the severity of symptoms. Studies show that adolescents who report higher levels of stress are more likely to experience intense mood disturbances and physical discomfort related to Premenstrual syndrome Steiner et al., (2006). This relationship is likely bidirectional, where premenstrual syndrome symptoms may increase stress levels, and stress can worsen premenstrual syndrome symptoms, leading to a vicious cycle. The pressures of academic performance, peer relationships, and societal expectations can amplify this stress, which is often compounded by a lack of coping mechanisms or emotional support Foley et al., (2008).

2.4.2. Lifestyle Factors

Lifestyle factors such as diet, physical activity, and sleep patterns play an essential role in the prevalence and severity of premenstrual syndrome symptoms among adolescent girls. A sedentary lifestyle, which is increasingly common among urban adolescents, has been shown to contribute to the development and worsening of premenstrual syndrome symptoms Sarkar et al., (2018). Lack of exercise not only reduces the overall fitness level but also limits the benefits of endorphin release, which can help alleviate mood disturbances and reduce stress. Physical inactivity is associated with higher levels of premenstrual syndrome-related symptoms such as fatigue, irritability, and depressive moods Blumenthal et al., (2007).

2.4.3. Dietary Influence

Dietary factors are another critical influence on the severity of premenstrual syndrome symptoms in adolescents. Research has suggested that diets high in salt, sugar, and caffeine can exacerbate premenstrual syndrome symptoms, while diets rich in fruits vegetables, and whole grains may help alleviate these symptoms Sarkar et al., (2018). Nutritional deficiencies, particularly low levels of magnesium, vitamin B6, and calcium, have been linked to the exacerbation of premenstrual syndrome symptoms, including irritability, mood swings, and cramps Pasquali et al., (2022). The study conducted by Oboza et. al., (2024) that low intake of simple carbohydrates, fats, salt, and alcohol, and high of fresh, unprocessed foods rich in B vitamins, vitamin D, zinc, calcium, and omega-3 fatty acids can help preventing the onset of PMS and reduce the severity of its symptoms in India, where dietary habits can vary greatly across regions, cultural dietary preferences may also play a role in the prevalence and intensity of premenstrual syndrome symptoms.

2.4.4. Sleep Patterns

Sleep disturbances are commonly reported among adolescents with premenstrual syndrome. Many adolescent girls experience poor sleep quality during the premenstrual phase, which can exacerbate the emotional and physical symptoms of premenstrual syndrome. Studies show that insufficient or disrupted sleep is associated with increased irritability, fatigue, and mood disturbances, all of which are characteristic of premenstrual syndrome Madan et al., (2021). Sleep deprivation can affect the hormonal balance, particularly the regulation of cortisol and melatonin, further intensifying Premenstrual syndrome symptoms. Establishing healthy sleep habits, such as regular sleep patterns and adequate sleep duration, may be beneficial in managing premenstrual syndrome.

2.4.5. Mood Disturbances

Mood disturbances, such as irritability, anxiety, and depression, are among the most common and debilitating symptoms of premenstrual syndrome in adolescents. These emotional symptoms can significantly impact daily life, including academic performance, social relationships, and overall quality of life. The hormonal fluctuations of premenstrual syndrome, combined with psychological stressors unique to adolescence, can create a perfect storm for mood-related issues Yonkers et al., (2008). Adolescents with more severe premenstrual syndrome may experience symptoms akin to premenstrual dysphoric disorder, a more severe form of premenstrual syndrome characterized by significant mood swings, irritability, and depression Freeman et al., (2011).

2.4.6. Exercise Adherence in Adolescents

Exercise has been shown to be an effective intervention for managing premenstrual syndrome symptoms, as it promotes the release of endorphins, improves sleep, and reduces stress. However, exercise adherence among adolescents is often low due to various factors, including lack of motivation, time constraints, and peer influence Blumenthal et al., (2007). Encouraging regular physical activity, such as aerobic exercise, could potentially alleviate many of the physical and emotional symptoms of premenstrual syndrome. However, promoting long-term exercise adherence in adolescents remains a challenge, and more research is needed to understand how to overcome barriers to exercise in this age group Snyder et al., (2003).

2.5. Symptoms of Premenstrual Syndrome

The exact etiology of premenstrual syndrome is not entirely understood but is thought to involve hormonal fluctuations, neurotransmitter alterations, and individual sensitivity to these changes Yonkers et al., (2008). Below is a detailed exploration of the symptoms, their prevalence, and associated impacts.

2.5.1. Physical Symptoms

Physical symptoms are often the most prominent and noticeable aspect of Premenstrual syndrome, causing significant discomfort for many. One of the most frequently reported symptoms is bloating and fluid retention, characterized by swelling of the abdomen, face, or extremities. This occurs due to hormonal changes leading to water retention and is experienced by up to 75% of individuals with premenstrual syndrome Direkvand-Moghadam et al., (2014). Accompanying the bloating is often a sense of heaviness or discomfort that can disrupt daily activities. Another common symptom is breast tenderness (mastalgia), which results from increased progesterone levels during the luteal phase. About 65% of women report experiencing sensitivity or pain in their breasts, with symptoms typically peaking just before menstruation Kues et al., (2002).

Headaches, including premenstrual syndrome-related migraines, affect approximately 45% of sufferers and are triggered by the hormonal shifts, particularly a drop in estrogen levels during the luteal phase MacGregor (2008) & Allais et al., (2012). Many individuals also experience fatigue and low energy levels, which can be linked to sleep disturbances and hormonal changes Shechter et al., (2012). These symptoms are further compounded by muscle and joint pain, often localized in the lower back or limbs, affecting about 60% of individuals with premenstrual syndrome Yonkers et al., (2008).

Additionally, digestive changes are commonly reported, including nausea, constipation, or diarrhea, due to prostaglandin activity and hormonal impacts on gut motility Brotto et al., (2015). Sleep disturbances such as insomnia, hypersomnia, or poor sleep quality also are prevalent and often exacerbate other physical and emotional symptoms Baker et al., (2012).

2.5.2. Emotional Symptoms

Premenstrual syndrome can significantly impact emotional well-being, with symptoms often mirroring or worsening pre-existing mental health conditions. Mood swings are among the most commonly recognizable emotional symptoms, marked by rapid and unpredictable changes in emotional states. Up to 85% of individuals with premenstrual syndrome report episodes of sudden sadness, anger, or irritability without clear provocation Friedman et al., (2019). Anxiety and tension are also prevalent, with heightened nervousness and difficulty in relaxing affecting 60–70% of cases Direkvand-Moghadam et al., (2014).

Depression presents as persistent feelings of sadness, worthlessness, or hopelessness. These symptoms are particularly common in severe Premenstrual syndrome cases and often overlap with PMDD, affecting about 30% of individuals Yonkers et al., (2008). Irritability and anger are also notable, often resulting in conflicts within personal or professional relationships due to heightened sensitivity to stressors.

2.5.3. Behavioral Symptoms

Behavioral changes associated with premenstrual syndrome can interfere with daily functioning and interpersonal relationships. One common symptom is difficulty concentrating, often described as "brain fog," where cognitive tasks such as decision-making or studying become challenging Farasati et al., (2015). Changes in appetite are also notable, with many individuals experiencing cravings for sweet, salty, or carbohydrate-rich foods. These cravings are thought to be driven by serotonin fluctuations and hormonal shifts, though some report a reduced appetite instead Brotto et al., (2015).

Social withdrawal is another key behavioral symptom, where individuals may avoid interactions or responsibilities due to physical discomfort or emotional distress Foley et al., (2018). This withdrawal can further exacerbate feelings of isolation and emotional instability.

2.5.4 Comorbidities and Overlapping Conditions

Premenstrual syndrome often co-occurs with other health conditions, complicating diagnosis and management. For instance, dysmenorrhea (painful menstruation) frequently overlaps with Premenstrual syndrome, with 64% of individuals having dysmenorrhea also reporting premenstrual symptoms. Shared symptoms include fatigue, abdominal pain, and irritability, amplifying the overall burden Iacovides et al., (2015). Mental health conditions, such as anxiety disorders and depression, are also more prevalent among individuals with severe premenstrual syndrome or PMDD Yonkers & Simoni, (2018). Furthermore, conditions like endometriosis, which share symptoms such as pelvic pain, often lead to diagnostic confusion and delayed treatment Biggs & Demuth, (2011).

2.6. Interventions for Premenstrual Syndrome

2.6.1. Pharmacological Interventions for Premenstrual Syndrome

Pharmacological interventions are commonly used to manage the symptoms of Premenstrual Syndrome, particularly when non-pharmacological treatments such as lifestyle modifications and exercise are not sufficient. The most commonly prescribed medications for premenstrual syndrome include non-steroidal anti-inflammatory drugs, anti-depressants, hormonal therapies and diuretics. Nonsteroidal anti-inflammatory drugs such as ibuprofen and naproxen, are frequently used to alleviate the physical symptoms of premenstrual syndrome, including cramps, headaches, and back pain. These medications work by reducing inflammation and blocking the production of prostaglandins, chemicals that cause pain and inflammation. A study by Garratt et al., (2006) demonstrated that Non-Steroidal Anti-inflammatory drugs were effective in reducing premenstrual syndrome-related pain, including dysmenorrhea (painful menstruation), with significant improvement in overall symptom relief. More recent studies Ravi Kumar et al., (2015), have reinforced the role of non- steroidal anti-inflammatory drugs in the effective management of physical premenstrual syndrome symptoms.

Selective Serotonin Reuptake Inhibitors (SSRIs), a class of antidepressants, are commonly prescribed for the mood-related symptoms of premenstrual syndrome, such as irritability, depression, and anxiety. SSRIs help increase serotonin levels in the brain, which can improve mood and emotional regulation. Freeman et al., (2013) found that SSRIs significantly reduced the psychological symptoms of premenstrual syndrome, particularly in women with severe mood disturbances and premenstrual dysphoric disorder. A study by Martín-Santos et al., (2017) further confirmed the efficacy of SSRIs, specifically fluoxetine, in alleviating both emotional and physical premenstrual syndrome symptoms.

2.6.2. Hormonal Therapies

This include oral contraceptives and hormone replacement therapy, aim to regulate the hormonal fluctuations that contribute to premenstrual syndrome symptoms. These therapies stabilize hormone levels, preventing the severe emotional and physical symptoms associated with Premenstrual syndrome. Wittchen et al., (2002) found that oral contraceptives were effective in reducing both emotional and physical premenstrual syndrome symptoms by stabilizing estrogen and progesterone levels during the menstrual cycle. Recent studies by Pothiawala et al., (2020) have demonstrated that newer hormonal therapies, including combined oral contraceptives with low-dose estrogen, are effective in significantly reducing premenstrual syndrome-related discomfort.

Diuretics, such as spironolactone, are sometimes prescribed to reduce water retention and bloating, common symptoms of Premenstrual syndrome. These medications work by increasing urine output, which helps reduce swelling and discomfort caused by fluid retention. Halbreich & Kuehner (2007) noted that diuretics are particularly helpful in addressing the physical symptoms of premenstrual syndrome, especially in women who experience significant swelling and water retention. More recently, Maguire et al., (2014) found that spironolactone effectively reduced symptoms of fluid retention, especially in women with severe premenstrual syndrome.

The use of these pharmacological interventions has been shown to significantly reduce the severity of premenstrual syndrome symptoms, particularly in women with moderate to severe cases. However, pharmacological treatments may come with side effects, such as gastrointestinal issues, mood changes, and weight gain, which need to be carefully considered when choosing the appropriate intervention. A study by Yonkers et al., (2008) found that SSRIs were particularly effective in reducing the psychological symptoms of premenstrual syndrome, including irritability and mood swings, while non-steroidal anti-inflammatory drugs provided relief from the physical pain associated with premenstrual syndrome.

2.6.3. Aerobic Exercise

Aerobic exercise is a documented intervention for reducing premenstrual syndrome symptoms. It helps to regulate hormones, reduce anxiety and depression, and improve overall mood by releasing endorphins. Aerobic exercise includes activities like running, cycling, and dancing, which increase heart rate and improve circulation, thus reducing bloating, fatigue, and mood swings. Running is a physical activity that involves moving swiftly on foot, typically over longer distances, and is known to elevate the heart rate and improve cardiovascular health. Cycling refers to riding a bicycle, a rhythmic exercise that works the lower body muscles while increasing heart rate and promoting cardiovascular endurance. Dancing involves rhythmic body movement, often to music, and can vary in intensity from light to vigorous, helping to improve coordination, strength, and cardiovascular function. All three activities engage large muscle groups and promote better circulation, which helps alleviate physical symptoms of premenstrual syndrome such as bloating and fatigue, while also improving mood and emotional well-being through the release of endorphins Koşar et al., (2021) found that regular participation in aerobic activities led to a decrease in the severity of both emotional and physical premenstrual syndrome symptoms. Additionally, exercise has been shown to increase serotonin levels, which can help alleviate the mood disturbances commonly associated with premenstrual syndrome. The study by Borenstein et al., (2015) examined the effects of structured aerobic exercise programs on premenstrual syndrome symptoms. It found that these programs significantly reduced premenstrual syndrome-related pain and mood disturbances by promoting emotional regulation and enhancing self-awareness among participants. The exercise routines, which included activities such as brisk walking, jogging, and cycling, were associated with improvements in mood, reduced pain levels, and better stress management. This research highlighted the role of physical activity in not only alleviating physical symptoms like bloating and discomfort but also in managing emotional symptoms such as irritability and anxiety, which are common in premenstrual syndrome. For instance, Daley (2009) found that women engaging in moderate-intensity aerobic exercise experienced notable reductions in premenstrual syndrome-related pain, fatigue, and emotional disturbances. This reinforces the role of exercise in not only managing immediate symptoms but also promoting overall health.

Similarly, Koşar et al., (2021) observed that university students participating in a structured aerobic exercise program reported a substantial decrease in the severity of premenstrual syndrome symptoms. The study highlighted improvements in physical discomforts like bloating and fatigue, as well as emotional challenges such as mood swings and irritability.

Furthermore, research by Dehnavi et al., (2018) emphasized the psychological benefits of aerobic exercise, including reductions in anxiety and depressive symptoms commonly associated with premenstrual syndrome. The study linked these effects to the release of neurotransmitters such as serotonin, which enhance mood and support long-term symptom management. Together, these findings underscore the holistic benefits of aerobic exercise in improving the quality of life for individuals with premenstrual syndrome.

2.6.4. Swimming

Swimming provides a full-body workout that is gentle on the joints while also offering a cardiovascular benefit. It is particularly beneficial for women with premenstrual syndrome because it reduces stress, improves circulation, and helps alleviate symptoms such as bloating and abdominal discomfort. The rhythmic movements of swimming combined with the buoyancy of water reduce muscle tension and promote relaxation. The study by Sundelin et al., (2019) investigated the impact of swimming on premenstrual syndrome symptoms among women, finding significant reductions in both psychological and physical symptoms. Participants who engaged in swimming experienced relief from common premenstrual syndrome symptoms like abdominal pain, irritability, and fatigue. The low-impact nature of swimming was particularly beneficial, making it suitable for women with severe cramps or joint pain, as it provided physical and emotional relief. This study highlights the effectiveness of swimming as a therapeutic exercise for managing premenstrual syndrome, particularly due to its gentle impact on the body.

2.6.5. Mindfulness

The mindfulness involves paying deliberate attention to the present moment in a non-judgmental way, often through practices such as meditation, deep breathing, or mindful movement (e.g., yoga). For women with premenstrual syndrome, mindfulness training can help manage emotional symptoms such as irritability, anxiety, and mood swings by promoting emotional regulation. Zhou et al., (2020) found that mindfulness-based interventions were effective in reducing the severity of premenstrual syndrome-related symptoms, particularly those related to mood. Mindfulness helps individuals cultivate awareness of bodily sensations, thoughts, and feelings, allowing them to respond to premenstrual syndrome symptoms with greater calmness and resilience. Mindfulness interventions have been shown to improve quality of life and reduce perceived stress, which plays a critical role in managing premenstrual syndrome.

2.6.6. Lifestyle Change

Lifestyle modifications are often recommended as part of a comprehensive approach to premenstrual syndrome management. These changes focus on improving diet, increasing physical activity, managing stress, and promoting better sleep hygiene. The study by Sarkar et al., (2018) explored how lifestyle changes, such as diet modification, could significantly impact premenstrual syndrome symptoms. The research involved a diverse group of women across different socio-economic backgrounds, including urban, urban slum, and rural areas in India. The study noted that reducing the intake of caffeine, alcohol, and processed foods while increasing whole grains, fruits, vegetables, and lean proteins helped balance hormone levels and improve overall well-being. The findings emphasized that these dietary changes could effectively reduce the intensity and frequency of premenstrual syndrome symptoms.

The study by Ussher et al., (2020) focused on the impact of stress reduction techniques like mindfulness, yoga, and relaxation exercises on premenstrual syndrome symptoms. This research was conducted among adolescent girls in a controlled setting, highlighting that these techniques were particularly effective in reducing mood swings and irritability often experienced during premenstrual syndrome. It showed that these techniques helped improve emotional regulation, making it easier for participants to manage premenstrual syndrome symptoms.

2.6.7. Cognitive Behavioral Therapy

Adolescence is a critical period of psychological and emotional development, making it essential to address the impact of PMS effectively during these formative years. Among various therapeutic approaches, cognitive behavioral therapy has emerged as a particularly promising intervention for adolescent girls dealing with premenstrual syndrome. Cognitive behavioral therapy offers targeted strategies to alleviate the psychological burden of premenstrual syndrome by addressing the interplay of thoughts, emotions, and behaviors.

Cognitive behavioral therapy has emerged as a highly effective intervention for managing premenstrual syndrome, particularly for adolescent girls. This structured, time-limited therapeutic approach addresses the psychological aspects of premenstrual syndrome by targeting maladaptive thought patterns and behaviors. Studies, such as those by Ussher et al., (2006) and Borenstein et al., (2007), indicate that cognitive behavioral therapy helps reduce premenstrual syndrome symptoms by improving mood, decreasing anxiety, and addressing negative automatic thoughts. It achieves these outcomes through psychoeducation, cognitive restructuring, and behavioral activation, fostering emotional regulation and effective coping mechanisms during critical developmental years.

The premise of cognitive behavioral therapy is rooted in the understanding that cognitive distortions, such as catastrophic thinking or overgeneralization, exacerbate emotional distress and premenstrual syndrome symptoms. Biggs et al., (2011) emphasize that cognitive behavioral therapy empowers individuals to identify and challenge these distortions, promoting healthier thought patterns and adaptive behaviors. This makes it particularly valuable for adolescents, as it not only mitigates immediate symptoms but also equips them with lifelong skills to manage stress and emotional challenges effectively.

2.6.8. Group Cognitive Behavioral Therapy

Group cognitive behavioral therapy is an evidence-based psychological intervention that has shown positive outcomes for women with premenstrual syndrome. It involves cognitive restructuring techniques that help participants identify and challenge negative thought patterns, reducing stress and improving emotional responses. The group dynamic adds an element of social support, which can be particularly valuable in managing emotional symptoms like anxiety and depression. A study by Ho et al., (2019) found that women who participated in group cognitive behavioral therapy reported lower levels of premenstrual syndrome-related anxiety, depression, and irritability. The study by Pistella et al., (2020) examined the role of group cognitive behavioral therapy in managing premenstrual syndrome symptoms, focusing on the social support aspect of the therapy. This research found that women who participated in group cognitive behavioral therapy reported lower levels of anxiety, depression, and irritability related to premenstrual syndrome. The group setting allowed participants to share experiences, coping strategies, and emotional support, which fostered a sense of solidarity among the women and enhanced the overall treatment effectiveness.

2.6.9. Multidisciplinary Approach

A multidisciplinary approach to premenstrual syndrome involves integrating multiple treatment modalities, including pharmacological treatments, psychological therapies, lifestyle adjustments, and physical interventions. By combining approaches from different disciplines, such as gynecology, psychiatry, and physical therapy, this approach targets the physiological, psychological, and social aspects of premenstrual syndrome. Studies indicate that combining therapies such as CBT with regular physical activity or dietary modifications can significantly improve symptom management. For example, the study by Miller et al., (2016) investigated the integration of mindfulness techniques with exercise as a method for alleviating both the emotional and physical burdens of premenstrual syndrome. It was found that combining these approaches was more effective than relying on a single treatment modality, particularly for those with moderate to severe symptoms. This combination helps manage stress and improves emotional regulation by enhancing mindfulness and exercise-induced endorphin release. The findings emphasize the importance of holistic approaches to managing premenstrual syndrome symptoms.

Additionally, the study by Barker et al., (2018) highlighted how such integrated approaches could significantly improve symptom management, reducing the frequency and severity of premenstrual syndrome-related symptoms. This research underscores the potential benefits of combining behavioral and psychological interventions with lifestyle modifications to enhance overall well-being and quality of life during premenstrual syndrome.

2.6.10. Yoga

Yoga is a mind-body practice that combines physical postures, breathing techniques, and meditation to promote overall health and well-being. It is particularly beneficial for women with PMS as it helps alleviate both physical and emotional symptoms. Yoga poses designed to stretch and relax muscles can help reduce abdominal cramps, lower back pain, and bloating, while the breathing exercises promote relaxation and stress relief. Additionally, yoga helps balance the nervous system, which can reduce symptoms like irritability, anxiety, and mood swings associated with premenstrual syndrome. A systematic review and meta-analysis by Majeed & Ahmad (2020) reinforced the effectiveness of yoga in reducing both physical and psychological symptoms of premenstrual syndrome, particularly pain, irritability, and fatigue. Their findings confirm that yoga's therapeutic benefits extend to both emotional and physical relief, making it a highly effective non-pharmacological intervention for premenstrual syndrome. Yoga's combination of gentle physical activity and mindfulness techniques provides a holistic approach to symptom management.

2.6.11. Aroma Therapy

Aroma therapy utilizes essential oils extracted from plants to improve physical and emotional health. This complementary therapy is particularly effective for managing the psychological symptoms of premenstrual syndrome, including mood swings, anxiety, and irritability, as well as physical symptoms like headaches and muscle tension. The soothing properties of essential oils such as lavender, clary sage, and chamomile can promote relaxation, improve sleep quality, and reduce discomfort during PMS. Kim & Lee, (2017) conducted a systematic review and meta-analysis, highlighting the significant benefits of aromatherapy in reducing premenstrual syndrome symptoms. Their analysis revealed that essential oils, particularly lavender and clary sage, were effective in alleviating mood swings, fatigue, and anxiety, making aromatherapy an accessible and non-invasive treatment for premenstrual syndrome. The calming effect of aroma therapy helps regulate mood and reduce stress, offering a natural solution to symptom management.

2.6.12. Acupuncture

Acupuncture, a form of traditional Chinese medicine, involves the insertion of thin needles into specific points on the body to stimulate energy flow (Qi) and restore balance. It has been used to treat a variety of conditions, including PMS. Acupuncture is thought to help regulate the menstrual cycle, reduce pain, and improve emotional well-being by enhancing blood flow and releasing endorphins, the body's natural painkillers. Zhao et al., (2018) conducted a systematic review of randomized controlled trials to assess effectiveness of acupuncture in managing PMS. Their findings supported acupuncture as an effective treatment, showing improvements in both physical symptoms, such as bloating and abdominal pain, and psychological symptoms, including irritability and mood swings. Ability of acupuncture to reduce pain and improve emotional stability makes it a promising non-pharmacological intervention for premenstrual syndrome.

2.6.13. Dietary Supplements

Certain dietary supplements have been found to alleviate premenstrual syndrome symptoms by addressing nutrient deficiencies or modulating hormonal levels. Commonly used supplements for Premenstrual syndrome management include calcium, magnesium, and vitamin B6. Calcium plays a role in reducing the severity of premenstrual syndrome symptoms like mood swings, bloating, and fatigue. Magnesium helps to alleviate cramps, reduce water retention, and improve overall mood stability. Vitamin B6 is involved in serotonin production and can help improve mood and reduce irritability. A study by Choi et al., (2016) highlighted the efficacy of vitamin D supplementation in managing Premenstrual syndrome, specifically targeting mood swings and fatigue by modulating calcium levels and improving serotonin function. Bajalan et al., (2019) found that calcium and magnesium supplementation led to a significant reduction in premenstrual syndrome symptom severity, especially bloating and mood swings. These findings suggest that nutritional interventions can play a key role in managing premenstrual syndrome, particularly for those with significant distress during their menstrual cycle.

2.6.14. Homeopathy

Homeopathy is a complementary therapy that stimulate the body's healing processes. Specific homeopathic remedies, such as Pulsatilla, Sepia, and Lachesis, are tailored to individual PMS symptoms, offering a personalized treatment approach. For example, Pulsatilla is used for emotional symptoms like mood swings and weepiness, while Sepia is indicated for irritability and fatigue. Berman & Cummings (2015) evaluated the effectiveness of homeopathy in managing premenstrual syndrome through a clinical trial and found significant improvements in both physical and emotional symptoms. Women treated with homeopathic remedies reported reduced pain, irritability, and fatigue, demonstrating homeopathy's potential in providing personalized symptom management for PMS. Its individualized approach ensures that treatment is specifically tailored to the unique needs of each patient.

2.6.14. Endocrinological Influence on Premenstrual syndrome

Premenstrual syndrome is a complex condition that is believed to be influenced by hormonal fluctuations in the menstrual cycle, particularly changes in estrogen and progesterone levels. These hormones play a central role in regulating the physiological processes associated with PMS. Endocrinological factors influence not only the onset of PMS symptoms but also their severity and duration.

2.6.14.1. Estrogen and Progesterone Fluctuations

The menstrual cycle is characterized by the fluctuating levels of estrogen and progesterone. Estrogen levels rise during the first half of the menstrual cycle and peak just before ovulation, while progesterone levels increase after ovulation and remain elevated during the luteal phase. For many women, symptoms of PMS emerge during the luteal phase, when progesterone levels are at their highest. Some research suggests that the interaction between these two hormones, particularly the relative imbalance between estrogen and progesterone, may contribute to PMS symptoms Yonkers et al., (2008).

The hypothesized mechanism behind premenstrual syndrome is that increased progesterone and the resultant increase in allopregnanolone, a neuroactive metabolite of progesterone can affect the central nervous system, including GABA (gamma-aminobutyric acid) receptors, leading to mood disturbances, irritability, anxiety and depression Roca et al., (2018). On the other hand, low estrogen levels in the luteal phase are believed to exacerbate physical symptoms such as bloating and fatigue, potentially due to its effects on serotonin and other neurotransmitter systems involved in mood regulation.

2.6.14.2. Role of Prolactin and Cortisol

In addition to estrogen and progesterone, other hormones like prolactin and cortisol may also play a role in Premenstrual syndrome. Elevated levels of prolactin during the luteal phase can contribute to breast tenderness, one of the hallmark symptoms of premenstrual syndrome Bhatia et al., (2008). Prolactin has a direct effect on breast tissue, making it more sensitive to hormonal fluctuations.

Cortisol, primary stress hormone, also interacts with premenstrual syndrome symptoms. Studies have shown that women with premenstrual syndrome often exhibit elevated cortisol levels in the luteal phase, potentially contributing to the psychological symptoms like anxiety, mood swings, and irritability Gordon et al., (2019). The interplay between elevated cortisol and other hormonal changes can exacerbate both physical and emotional symptoms of PMS.

2.5.14.3. Thyroid Dysfunction

Thyroid dysfunction, particularly hypothyroidism, is another endocrine factor that can influence the severity of PMS symptoms. Women with thyroid imbalances may experience more severe PMS symptoms, particularly fatigue, weight gain, and mood disturbances. Studies have shown that thyroid hormone levels may influence the balance of estrogen and progesterone, thereby worsening PMS Sharma et al., (2015).

Conversely, hyperthyroidism has been linked to increased anxiety, irritability, and sleep disturbances, which can overlap with PMS like symptoms. Hormonal dysregulation involving the thyroid may therefore worsen or mimic premenstrual syndrome, making it difficult to differentiate the two conditions.

2.6.15. Cultural Perspectives

Cultural beliefs and norms significantly influence how PMS is perceived and addressed globally. In many cultures, PMS remains stigmatized or regarded as a taboo topic, which can deter women from seeking help or discussing their symptoms openly. This cultural silence often leads to under diagnosis and untreated cases, further exacerbating the challenges faced by affected individuals. Conversely, in some societies, premenstrual syndrome is openly acknowledged, with traditional remedies or community support systems frequently utilized to alleviate symptoms.

Understanding and integrating cultural context into healthcare practices can greatly enhance the effectiveness of interventions. Culturally sensitive education programs, as suggested by Foley et al., (2018), play a critical role in improving awareness and reducing stigma. These programs empower women by normalizing conversations about premenstrual syndrome and providing tailored information that resonates with their cultural values and beliefs. Such initiatives encourage women to seek professional assistance and adhere to treatment plans, fostering better outcomes.

Healthcare providers can also benefit from cultural competence training, which enables them to understand and respect diverse perspectives on PMS. By recognizing cultural differences in symptom expression, treatment preferences, and help-seeking behaviors, providers can deliver more personalized and effective care. Integrating traditional remedies with evidence-based practices, where appropriate, can further enhance treatment adherence and acceptance across varied populations.

2.6.16. Education

Education about premenstrual syndrome is critical for both individuals experiencing symptoms and healthcare professionals. Many women may not recognize their symptoms as part of a medical condition and may not seek treatment until their symptoms become unmanageable. Public health campaigns, school programs, and healthcare provider training can increase awareness of PMS and improve its diagnosis and management Miller et al., (2017). Educating women about lifestyle modifications, the role of exercise, and effective stress management strategies empowers them to take an active role in managing their health. Providing comprehensive education can also reduce the stigma surrounding PMS, helping women feel more comfortable discussing their symptoms with healthcare professionals and peers.

2.6.17 Peer Support

Peer support can be a valuable resource for women experiencing premenstrual syndrome. Informal support of networks with family, friends or support groups provide emotional comfort and help individuals share coping strategies. Peer support can reduce feelings of isolation and validate the experiences of women with premenstrual syndrome, encouraging them to take active steps in managing their symptoms.

The study by Karami et al., (2020) explored the impact of peer support groups and community-based programs on emotional well-being and symptom management among women experiencing PMS. The research found that these peer support groups significantly improved emotional outcomes, such as reducing anxiety, depression, and irritability, by providing a sense of community and shared understanding among participants. Women who engaged in these support networks reported better management of symptoms and greater emotional stability compared to those who did not participate. This emphasizes the value of peer support in enhancing the psychological aspects of premenstrual syndrome treatment, offering a complementary approach to more formal therapeutic interventions.

2.6.18. Comparative efficacy of Cognitive Behavioral Therapy and Aerobic Exercise in Premenstrual Syndrome Management.

While both cognitive behavioral therapy and aerobic exercises have demonstrated efficacy in managing premenstrual syndrome symptoms, their mechanisms of action differ. Cognitive behavioral therapy primarily addresses the cognitive and emotional dimensions, whereas aerobic exercise exerts its effects through physiological pathways. A comparative evaluation is essential to determine their relative effectiveness, especially in adolescent populations where accessibility, adherence, and preference may vary.

Research suggests that combining interventions, such as aerobic exercise and CBT may provide synergistic benefits for managing premenstrual syndrome. Physical activity not only alleviates physical symptoms but also enhances cognitive function and mood, potentially amplifying the effectiveness of cognitive behavioral therapy. Koşar et al., (2021), highlighted that regular aerobic exercise improves emotional regulation, which could complement cognitive behavioral therapy focus on addressing maladaptive thought patterns and behaviors.

These promising indications, direct comparative studies assessing the combined effects of exercise and CBT are limited. Further research is needed to explore how these interventions interact and whether their integration leads to greater improvements in premenstrual syndrome symptom management compared to either approach alone. This integrated strategy could represent a holistic and effective solution, particularly for adolescents and young women experiencing significant premenstrual syndrome symptoms.

2.7. Barriers to Treatment

Despite the availability of effective treatments for PMS, numerous barriers hinder women from accessing proper care. These obstacles include a lack of awareness about available treatment options, cultural stigmas surrounding menstrual health, financial constraints, and limited access to healthcare providers. Many women with premenstrual syndrome symptoms fail to seek treatment due to social stigma, embarrassment, or the belief that their symptoms are a normal part of womanhood, often overlooking the possibility of effective interventions Bajalan et al., (2019).

2.7.1. Awareness and Education Gaps

A significant barrier to seeking treatment is the lack of awareness about premenstrual syndrome as a treatable medical condition. Many women do not recognize the severity of their symptoms or may dismiss them as normal cyclical discomfort. According to a study by Roberts et al., (2017), many women experience premenstrual syndrome for years without seeking medical advice because they consider the symptoms to be part of the natural menstrual cycle. Increased awareness about premenstrual syndrome and its potential treatments can empower women to seek medical help earlier.

2.7.2. Cultural Stigma and Social Norms

Cultural factors also play a pivotal role in the management of premenstrual syndrome. In many societies, menstrual health is a topic surrounded by stigma and silence. Cultural taboos may prevent women from discussing their symptoms or seeking treatment, with many individuals accepting menstrual discomfort as an inevitable part of being female Liu et al., 2020). These cultural norms can result in women suffering in silence, delaying diagnosis and treatment. For example, in some cultures, menstruation and premenstrual syndrome are seen as taboo topics, and discussing them openly may be perceived as inappropriate or shameful Tian et al., (2018).

2.7.3. Financial Constraints

Financial barriers also limit access to care for premenstrual syndrome. According to a study by Smith et al., (2015), the cost of healthcare, including consultations and treatments for PMS, can be prohibitive for many women, particularly in low-income communities. This is exacerbated by the lack of insurance coverage for non-pharmacological treatments such as therapy or exercise programs that can alleviate PMS symptoms. Women in economically disadvantaged situations may delay or forgo treatment due to the associated costs, leading to worsened symptoms and decreased quality of life.

2.7.4. Limited Access to Healthcare Providers

Another barrier to treatment is the limited access to healthcare providers who specialize in premenstrual syndrome management. Women in rural or underserved areas face difficulties accessing gynecologists or other healthcare professionals who can properly diagnose and treat premenstrual syndrome. This issue is compounded by a shortage of trained specialists in some regions, which can lead to long waiting times for appointments or inadequate management of premenstrual syndrome symptoms Kaur et al., (2016). Furthermore, general practitioners may not always have the time or expertise to properly address the specific needs of women suffering from PMS.

2.7.5. Addressing Barriers to Treatment

In order to overcome these barriers, it is crucial to implement strategies that enhance awareness and accessibility. Education campaigns that address the physical and psychological aspects of premenstrual syndrome, reduce stigma, and inform women about available treatments can play a key role in promoting early intervention. Providing affordable and accessible healthcare options, particularly in rural or underserved areas, can also improve access to care for women who need treatment for premenstrual syndrome Miller et al., (2020). Furthermore, integrating mental health services into Premenstrual syndrome management can help address the emotional and psychological challenges faced by women dealing with premenstrual syndrome, improving overall outcomes.

3. Methodology

This study was conducted on adolescent medical students to evaluate the effect of cognitive behavior therapy and aerobic exercises in alieving premenstrual syndrome.

3.1. Study design: It is an interventional study

3.2. Study population: The study was conducted on students of constituent colleges of Adesh University, village Bhuccho Kalan, District Bathinda, Punjab, residing in the hostels of the Adesh University.

3.3. Pilot study

A pilot study was carried out to determine feasibility of the study in May 2021. Fourteen girl students of 1st semester of College of Physiotherapy, Adesh Institute of Medical Science and Research, Adesh University, Bathinda, were recruited as participants of the study. The participants were informed about the objectives of the study and methodology to be followed through an elaborate lecture. The participants were advised to fill daily record of severity of symptoms problem (DRSP) diary for one week. Out of the 14 girls taken for the pilot study, 04 girls were menstruating in following days, hence these 04 girls were dropped off and flagged so that they can be recruited 14 days before menstruation. The girls showing premenstrual symptoms on DRSP were advised to fill Premenstrual symptoms screening tool for adolescents (PSST–A) and Premenstrual syndrome scale (PMSS). Out of the remaining 10 girls, two (02) girls did not show any PMS symptoms, hence they were dropped off. The feasibility of the sample was 80%. In order to check the procedural understanding, remaining 8 girls were asked to fill pre intervention questionnaire. The participants were divided into 02 groups, one group was administrated one session of CBT and the second group was administrated one aerobic exercise session to determine the working of interventional protocols.

3.4. Sample size

A priori power analysis was performed, to determine the necessary sample size, using GPower 3.1 Kang, (2021) with pain disability as the primary outcome. The number of subjects to be involved in the study was calculated by considering the sample size, 0.5 effect size, confidence interval 95% and 80% power of the previous study of Weise (2019). Based on these assumptions the desired sample size was 126 (64 in group A and 64 in group B). In the present study, 144 adolescent girls were included to compensate up to a 15% dropout rate. Hence, 72 adolescent girls were included in both group A (CBT group) and group B (Aerobic exercise group). Each participant was assigned a unique ID to ensure confidentiality.

3.5. Inclusion criteria

- Adolescent girls between the age of 18-19 years studying in a constituent college of Adesh University, Bhucho Kalan, Bathinda, Punjab.
- Girls must have attained menarche and having regular menstruation.
- Girls with premenstrual symptoms in previous consecutive two months and recorded on daily record of severity of symptoms problem (DRSP)
- Willing to participate

3.6. Exclusion criteria

- Pre-diagnosed case of hypothyroidism, asthma, epilepsy, pelvic inflammatory disease, endometriosis, polycystic ovary syndrome, psychosis, bipolar disorder, somatic symptom disorder and acute suicidal tendencies.
- Participated / participating in psychotherapy for same cause in the past or currently.
- Have taken antidepressant/antipsychotic or hormone therapy in last 3 months.
- Did not maintain symptoms recorded on daily record of severity of symptoms problem diary (DRSP)

3.7. Selection protocol

The voluntarily participated girls were further scrutinized according to the guidelines of the American College of Obstetrics and Gynaecology and DSM-V. According to the guidelines the adolescent girls should fulfil following criteria to be diagnosed as PMS:

- Girl should express at least one core affective symptom out of the following:
    a) Depression
    b) Anger outburst
    c) Irritability
    d) Anxiety
    e) Confusion and social withdrawal
- At least one core somatic symptom should be present out of the following:
    a) Breast tenderness
    b) Abdominal bloating
    c) Headache
    d) Swelling of the extremities.
- These affective symptoms and somatic symptoms should be present in the luteal phase (4-6 days before start of bleeding) and disappear in follicular phase (5 days after post menses).
- The daily record of severity of symptoms problem (DRSP) diary should be maintained for at least 2 months.

3.8. Materials Used

3.8.1. Weighing Machine: A weighing machine (Hi Labs HCL9013) device was used to measure an individual's body weight. It provided readings in kilograms (kg) and it was considered essential for monitoring weight-related changes in clinical, research, or personal settings Centres for Disease Control and Prevention, (2020).

3.8.2. Stature Meter: A stature meter (Sahyog Wellness Human Body Stature meter) is an instrument used to measure height of an individual accurately. A stature meter typically consists of a vertical measuring scale and a movable headpiece that is positioned at the top of the head of individual. Measurements were recorded in centimeters (cm), ensuring precision for anthropometric assessments Centers for Disease Control and Prevention, (2020).

3.8.3. Sociodemographic Data Sheet: comprised of following question:

i. The sociodemographic data sheet (Appendix-I) consists of twenty-one questions divided into three sections given below:

ii. **Part 1** covered participants' personal details, including current age, age at menarche, religion, family type, annual income, residential area and address, parental education level, and employment status.

iii. **Part 2** focused on body composition, including body measurements of height in centimeter and weight in kilograms. Body Mass Index (BMI) was calculated as weight/height^2.

iv. **Part 3** addressed menstrual details, covering aspects such as menstrual cycle regularity, cycle duration, flow duration (in days), flow nature, personal hygiene practices, type of absorbent used, and frequency of absorbent changes.

3.8.4. Daily Record of Severity of Symptoms Problem Diary: The Daily Record of Severity of Symptoms Problem Diary (DRSPD) is a standardized self-report tool designed to track the daily symptoms of premenstrual syndrome (PMS) or premenstrual dysphoric disorder (PMDD) over a menstrual cycle. It includes a series of items rated on a 6-point Likert Scale, ranging from 0 (not at all) to 5 (extreme), assessing physical, emotional, and behavioural symptoms. The diary aids in identifying symptom pattern and their severity, facilitating diagnosis and evaluation of treatment effectiveness. The DRSP is a reliable instrument (Appendix-II) with strong validity for tracking premenstrual symptoms Endicott and Harrison (2006).

3.8.5. Premenstrual Syndrome Screening Tool for Adolescent: The Premenstrual Syndrome Screening Tool for Adolescents (PSST-A) is a questionnaire specifically designed to identify and assess PMS in younger populations. It includes questions focused on the physical, emotional, and behavioural symptoms commonly associated with PMS, as well as their impact on daily activities, school, and social interactions. Adolescents rate the presence and severity of symptoms over a typical menstrual cycle, providing insights for diagnosis and management. The tool is validated for use in adolescent populations, demonstrating good reliability and construct validity. It is a revised scale specially designed for adolescent girls to diagnose moderate to severe PMS (Appendix –III).

A participant should present with at least one of the symptoms listed as number 1, 2, 3, and 4, in addition to this at least four out of 1-14 symptoms should be present Steiner et al., (2011).

3.8.6. Premenstrual Syndrome Scale (PMSS): The impact of premenstrual symptoms was assessed by premenstrual syndrome scale. The premenstrual syndrome scale comprised of 40 questions with three sub-scales, physiological, psychological and behavioral symptoms. These questions were answered on a 5-point Likert Scale. Responses were: Never -01 point; Rarely – 02 point; Sometimes – 03 point; Very often – 04 points and Always - 05 points. Interpretation of the PMSS was done on percentage. The total score obtained from the sub-scales was termed as PMSS total score. On the PMS scale the lowest score is 40 and the highest score is 200. If the PMS scale total score reached 80 points or above, this indicated the occurrence of PMS. Increases in the scores indicate an increase in PMS severity. Based on the percentage of scores, the levels of premenstrual symptoms were graded in following four categories:

No symptoms; mild symptoms; moderate symptoms; severe symptoms and very severe symptoms Padmavathi and Dhanapal (2014) (Appendix–IV)

3.8.7. Pain Disability Index: This pain disability index (PDI) is a self-reported questionnaire designed to assess the level of disability caused due to pain in daily life. It consists of 7 items, each item was scored on a 0–10 Likert Scale, where; 0 represented no disability and 10 indicates total disability. The domains evaluated included family/home responsibilities, recreation, social activity, occupation, sexual behavior, self-care, and life-support activities. The total score was obtained by summing the scores of all items, with higher scores indicating greater disability due to pain. The PDI demonstrates (Appendix–V) good psychometric properties, including high internal consistency, with reported Cronbach's alpha values typically exceeding 0.85, indicating excellent reliability. It also exhibits strong construct and criterion validity, making it a reliable tool for assessing pain-related disability in clinical and research settings Taitat et al., (1990).

3.8.8. Health Related Short Form-36 Questionnaire (RAND): The Health-Related Short Form-36 Questionnaire (RAND Sf 36) was used as self-administered tool designed to evaluate health-related quality of life (HRQoL) across eight domains: physical functioning, role limitations due to physical health, role limitations due to emotional problems, energy/fatigue, emotional well-being, social functioning, pain, general health and health change. Each domain was scored on a scale from 0 to 100, where higher scores indicated better health status. This questionnaire (Appendix–VI) is well-validated, demonstrating strong psychometric properties, including high internal consistency and excellent construct and criterion validity. It is widely used in both clinical and research settings to assess and monitor changes in health-related quality of life over time Wareand Sherbourne, (1992).

3.8.9. Beck Depression Inventory-II: The Beck Depression Inventory-II (BDI-II) is a widely used self-report questionnaire designed to assess the severity of depressive symptoms. It consists of 21 items, each scored on a 4-point Likert Scale ranging from 0 to 3, reflecting the intensity of specific symptoms experienced over the past two weeks. The items cover a range of emotional, cognitive, behavioral, and physical symptoms of depression, including sadness, guilt, loss of pleasure, fatigue, and changes in appetite or sleep patterns. The total score was obtained by summing the scores for all items, with higher scores indicating greater severity of depression. The BDI-II (Appendix-VII) categorizes depression severity into minimal, mild, moderate, and severe based on specific score thresholds. The instrument demonstrates excellent psychometric properties, with good internal consistency and reliability. It also has robust construct and criterion validity, making it a reliable and effective tool for both clinical practice and research in evaluating depressive symptoms Becket et al., (1961).

3.8.10. Beck Anxiety Inventory: Beck Anxiety Inventory (BAI) (Appendix–VIII) comprises 21 self-rating items scored on a 4-point Likert Scale that ranged from 0 to 3. This scale assesses the impact of each anxiety symptom in the preceding week of menstruation. The final score for anxiety was obtained by summing the scores obtained for all the items and the final score was used to classify severity of the anxiety into three categories: Low, Moderate and High. The BAI demonstrated strong internal consistency. It also had good construct and criterion validity Beck et al., (1988).

3.9. Diagnostic assessment

- Retrospect record of symptoms for two consecutive cycles was maintained by daily record of severity of symptoms problem.
- Participants were screened for premenstrual syndrome with the help of premenstrual syndrome screening tool for adolescent (PSST-A).
- Premenstrual syndrome scale was used to measure the impact of symptoms of PMS.

3.10. Demographic variables

Demographic data was recorded in a structured format which included name, gender, education, age, home address, social economic status, occupation and parental income and education status of the parents. Anthropometric data on weight, height, body mass index, age when menstruation started (menarche) was also recorded. Personal hygiene such as type of material or absorbent used (cloth, cotton, homemade pads, commercial pads) during menstruation, time gap for changing material/absorbent during menstruation (4 hourly, 6 hourly and 10 hourly), cleaning of private parts during menstruation. Information about menstrual characteristics like menstrual cycle duration, perceived menstrual flow, perceived BMI after menarche, span of cycle etc. were also collected. In order to analysing data more precisely, a Likert scale was formed for age at menarche, present BMI, span of cycle, perceived BMI after menarche and duration of flow so that data can be interpreted. Age at menarche was having range of age from 10 to 15 years hence 10 -1;11-2; 12-3; 13-4; 14-5 and 15 -6. Present BMI was scaled as per WHO, <17.0 (thin)-0; <18.5(underweight)-1; 18.5-24.9 (normal weight) - 3; 25-29.9 (over weight) - 3 and 30>35.7 (obese) - 4. The span of cycle was ranged from 31 days to 24 days therefore Likert scale was 0 was for 24 days and 07 was for 31 days. The likert scale for perceived BMI after menarche was Very lean-0; lean -1; normal-2 and over weight- 3. Likert scale used for duration of flow which was ranging from 8 days to 3 days hence scale was from 5 to 3. 8 days -5; 07-4 days; 06- 3 days; 05-02; 4 days -01 and 3 days as 0.

3.10.1. Primary outcomes

- Pain Disability Index was used to evaluate the interference of chronic pain in daily activities (Tait at el., 1990).
- Health Related Short Form-36 questionnaire was used to assess the quality of life of the adolescents Ware and Sherbourne, (1992).

3.10.2. Secondary outcomes

- Anxiety in adolescent girls was evaluated by Beck Anxiety Inventory Beck et al., (1988).
- Depression in adolescent girls was evaluated by Beck Depression Inventory-II Beck et al., (1961).

3.11. Enrolment procedure

225 first year girl students, aged 18-19 years of various constituent colleges of Adesh University, Bathinda, Punjab were approached during academic years 2022-2023. An interactive lecture for students was conducted in each constituent college of the Adesh University during the student induction program and the willing candidates who wanted to participate in the study were recruited. Twenty-three girls were excluded as they were not residing in the hostel. The remaining 202 girls residing in the university hostels were included. In this study voluntaries belonged to College of Nursing (75), College of Physiotherapy (55), Adesh Institute of Dental Sciences (30) College of Para Medical Sciences (28) and Adesh Institute of Pharmacy (06).

Written consent was taken from each volunteer after explaining the objectives of study and the procedure to be followed (Appendix-IV). All participants were asked to fill sociodemographic data. According to the guidelines of the American College of Obstetrics and Gynaecology and DSM-V, five (05) girls were excluded as they did not fulfil the eligibility criteria due to irregular menstruation. The remaining 197 girls fulfilling the inclusion criteria were asked to fill DRSP diary. All the terms of the DRSP diary was explained to the participants and given explicit instructions to remember the symptoms and complete the daily for two months and it was deemed complete if all the days were marked. The completion of the DRSP dairy was reminded frequently. Every month, DRSP dairy was collected, and every column was examined. Only fully complete dairies were collected and used for study. Three students (03) were eliminated because they were unable to finish the DRSP diary. 194 girls completed DRSP dairy after which premenstrual symptoms screening tool was used to diagnose premenstrual syndrome. 37 girls did not meet the criteria of PSST-A, hence, they were also excluded from the study. The remaining 157 participants were instructed to fill PMSS to check the extent severity of the symptoms. Of these, 154 participants showed moderate PMS and three girls showed mild symptoms. Thirteen (13) girls were dropped from the study due to reasons like leaving hostel, could not adhere to sessions due to examinations and vacation for eight weeks. Total 144 girls completed sessions of CBT and aerobic exercises.

These 144 participants of the study were randomly divided into two groups i.e. Group A (CBT group) and Group B (Aerobic exercises group) by an independent assessor. Each participant was given a unique ID and they were randomized to constitute two groups, using simple randomization method. Flip coin method was used to eliminate selection bias and to ensure that the groups were comparable. An independent researcher, who was not involved in the study, performed the coin flips and recorded the group assignments. Each group was consisted of 72 participants. One side of the coin was assigned to each of the groups. The coin head was given to group A and tails was given to other the group B. The coin was flipped for each participant individually and after each coin flip, the participant was assigned to the group corresponding to the flipping outcome and the same was documented for every participant. Record sheets of group assignment were sealed in the password protected electronic file and only ID numbers were used throughout the study instead of names of the participant. The coin was flipped till both groups were having 72 volunteers. All eligible participants underwent following assessment through the respective questionnaire.

- Pain Disability Index Questionnaire (Tait et al., 1990)
- Health Related Short Form 36 questionnaire (Ware & Sherbourne, 1992)
- Beck Depression Inventory (Beck et al., 1988)
- Beck Anxiety Inventory (Beck et al.,1961)

3.12. Interventions protocol

72 participants of Group A (CBT group) received Cognitive Behavioral Therapy. The CBT has been given in 4 steps.

Step 1: Identifying Negative Thoughts and Beliefs

a. Recognizing Negative Thought Patterns

Participants were made aware of negative thought patterns that could be causing or exacerbating emotional distress. Negative thought patterns often manifest as automatic, recurring thoughts that contribute to negative emotions and unhelpful behaviors

b. Challenging Negative Beliefs

After identification of negative thought patterns have been identified, the next step was to challenge and reframe these negative beliefs. This process was involved critical evaluation of the evidence supporting these beliefs and considering alternative, more balanced perspectives.

Step 2: Understanding the Connection Between Thoughts, Feelings, and Behaviors

- As per ABC model of CBT the individual's beliefs or thoughts about the event lead to negative emotional consequences, such as feelings of worthlessness, sadness, and withdrawal from colleagues.
- By implementing the ABC model and examining their own thought patterns, individuals can gain greater self-awareness and make positive strides towards achieving mental resilience.

Examining the Relationship Between Thoughts, Feelings, and Behaviors

Participants were encouraged to examine the relationship between their thoughts, feelings, and behaviors and explore this connection.

Participants were guided to:

1. **Identify Automatic Thoughts**: Automatic thoughts are the immediate, spontaneous thoughts that arise in response to a situation. These thoughts often occur without conscious awareness and could be negative or distorted. Recognizing these automatic thoughts was the first step in understanding their impact on emotions and behaviors.
2. **Challenge Unhelpful Beliefs**: Once automatic thoughts were identified, participants were encouraged to challenge and evaluate the accuracy and validity of these thoughts. This process involves examining evidence for and against the thoughts, considering alternative explanations, and adopting more realistic and balanced perspectives.

Step 3: Restructuring Negative Thoughts

This step was involved identifying and challenging negative thought patterns, as well as reframing those thoughts into more positive and balanced perspectives

1. **Identifying Cognitive Distortions**: By learning to identify and challenge distortions, Participants gained a clearer understanding of their thoughts and beliefs.
2. **Examining Evidence**: Participants were encouraged to examine the evidence for and against their negative thoughts which can help them gain a more

objective view of the situation. This step involved evaluating the facts which were gathered and considering alternative interpretations that were more realistic and balanced.

3. **Decatastrophizing**: By considering more realistic and less extreme outcomes of the thoughts. This technique helps in reducing anxiety of the participants and developed a more rational perspective of the "Hot thought".

Reframing Negative Thoughts

By reframing negative thoughts and consciously changing the way one perceived and interpreted negative events or situations. Participants could create a more positive and adaptive mindset.

Step 4: Developing Coping Strategies

The negative thoughts could be managed by developing effective coping strategies. By building coping skills and implementing healthy coping mechanisms, Participants could develop.

Following coping strategies were given to the participants

1. **Relaxation Techniques**: Deep breathing exercises, progressive muscle relaxation, and mindfulness meditation.
2. **Problem-Solving Skills**: Developing problem-solving skills, participants were taught breaking down of problems into smaller, manageable steps, so that participants can approach them more effectively.
3. **Social Support**: Building a support system of friends, family, or support groups can provide a sense of connection and understanding. Sharing experiences and receiving support from others could help in alleviate stress and promote resilience
4. **Healthy Lifestyle Habits**: Participants were encouraging to engage in regular physical exercise, to take balanced diet and self-care that contribute to overall well-being and resilience of the participants.

Implementing Healthy Coping Mechanisms

Participants were encourage to use coping mechanisms so that their negative thoughts could get outlet by expressing them without the fear of mistrust. Coping Mechanisms were Journaling, developing hobbies,practicing self-care activities and positive self-talk.Since it was not possible to handle all 72 participants in a session, subgroup of 8-10 participants was formed. Each subgroup was given one session per week for 8 weeks according to Maddineshat et al., (2016). The protocol is given in Table 3.12.1.

Table 3.12.1. Cognitive Behavioral Therapy Protocol

Session	Work to be accomplished
Week 1	Introduction of therapist and group members. Students were asked to fill the questionnaires. Therapist gave description of premenstrual symptoms; elimination of signs and symptoms; therapeutic agents; outcomes and impacts; the role of environmental stressors and psychological factors. **Assignment:** Filling up of Pain Disability Index, Short form 36 (Sf36), Beck Anxiety Inventory and Beck Depression Inventory.
Week 2	Description of group therapy, cognitive-behavioral interventions, duties and norms were informed to the subjects. In order to increase motivation and group cohesion; description of group objectives, description of group meeting and determining behavioral changes were explained. Initial anxiety of group members was encountered by opening up resistance and promoting self-disclosure. **Assignment**: Recorded moods one week before the start of menstrual cycle, and noted down physical changes caused by moods.
Week 3	Evaluation of automatic thought and cognitive distortions; emotional reasoning, generalized extreme, discussion and analysis of anxiety, depression, and symptoms that occur periodically. Reviewing the basis of specific experiences by retaining confidentiality and providing feedback. **Assignment**: Detect thoughts, moods and behaviors experienced in different life situations and identify which parts of those experiences need to be changed.

Week 4	Training of members on emotional reactions and their relation to internal dialogue; exploring situations; having a broader and more objective perspective; dealing with resistance, tension, and anxiety. **Assignment**: participants were asked to consider a recently experienced mood which impacted them, identify moods during or immediately after it; self-reflect, observe and classify mood swings.
Week 5	Explain, interpret, and describe the relationship between situation and emotion; awareness of situations and thoughts that relate to mood changes; describe thoughts, beliefs, concerns, perceptions, and concepts related to situations. **Assignment**: Note down support for the inaccuracy of "hot thoughts". Gather evidence to confirm or reject a hot thought.
Week 6	Evaluating thoughts and challenging them; role playing in the group to reform and challenge automatic thoughts; analyzing behavioral changes in various situations; assessing personal responsibility; identifying main beliefs. **Assignment**: Test new experiences, experiments and practical plans.
Week 7	Identifying underlying principles and assumptions; therapeutic interventions; identifying worries and their benefits and harms; investigating ways to manage negative events of the past; identifying and classifying violent thoughts and emotions associated with them; mental imagery. **Assignment:** Analyze new information and acceptance.
Week 8	Treatment of depression; cognitive restructuring; improving interpersonal relationships; timed schedule of activities; recording weekly activities; taking a lesson from these schedules; understanding anxiety, its characteristics and cognitive aspects; practicing relaxation, controlled breathing, visualization and distraction; understanding anger, guilt and

	shame; anger restraint methods; assertiveness training; overcoming feelings of guilt. **Assignment**: Interactive lecture to be delivered on menstrual hygiene.

Group B also consisting of 72 participants (Group Aerobic Exercise) received aerobic exercises for the duration of 50 minutes, three days in a week for eight weeks. It included three phases, warm up phase, aerobic exercises and cool down phase. Warm up and cool down phase were of 10 minutes and each movement for 16 counts Annapoorani & Lathavenkatesan, (2016). One session was given to participants one week before starting the actual exercise regime so that they can learn the movements. At the end of all 8 sessions an interactive lecture was delivered on menstrual hygiene.

a. Warm Up Phase: The warm up exercises given in table 3.13.1 were carried out for 10 minutes.

Table 3.13.1 Warm up protocol of aerobic exercises

Sr.No	Body Movement	Sr.No	Body Movement
1.	**Heel movements-** In standing position lift the alternate heels. 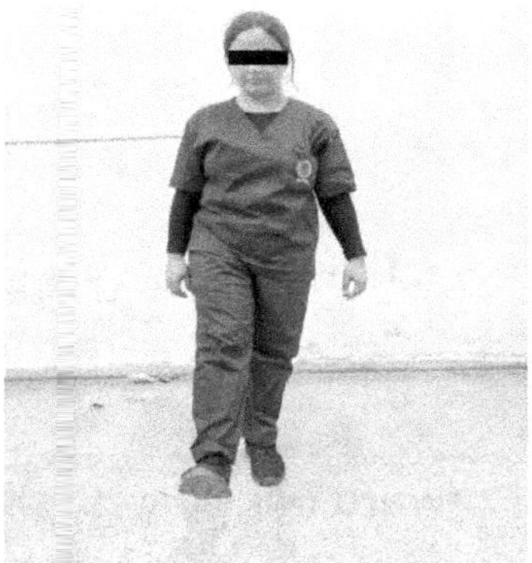	2.	**Upward reaches-** One step sideways on right side and one step sideways on left and front, hand up 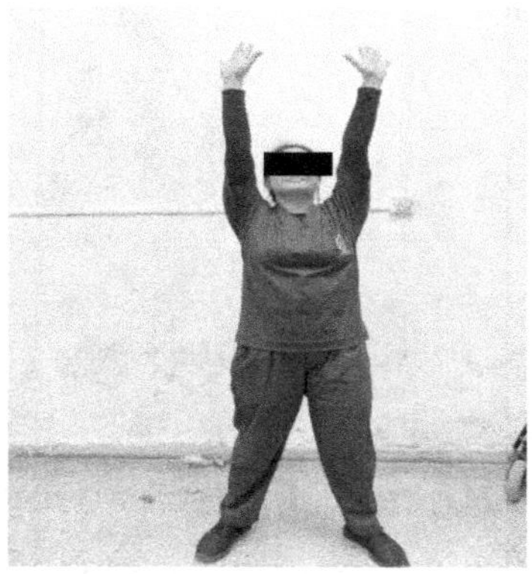
3.	**Neck movements-** Rotate neck clockwise and anticlockwise rotation. Flexion and extension of the neck 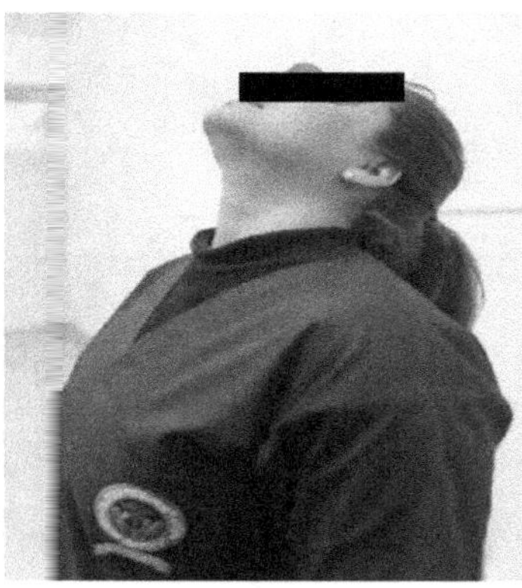	4.	**Arm rotation-** Stand with arms extended out to the sides at shoulder height, rotate both arm clockwise and anti-clockwise. 

Continued...

5.	**Trunk rotation**- Sit upright in a chair, feet flat on the floor, turn torso towards the right side and left side. 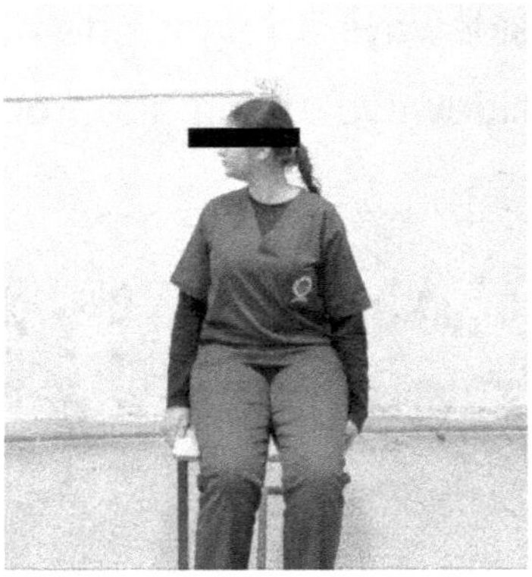	6.	**Hip rotation**- Stand tall, lift one knee to hip height and rotate it clockwise then rotate it anti clockwise. 
7.	**Knee rolling**- Bend forward bend knee and rotating in the clockwise and anticlockwise direction. 	8.	**Forward bending-** In sitting position bend your trunk and touch the feet with the hands. 

Continued...

9.	**Hamstring stretches**- Take a step forward; keep front leg straight bend forward by keeping back straight; keep the backward leg straight. 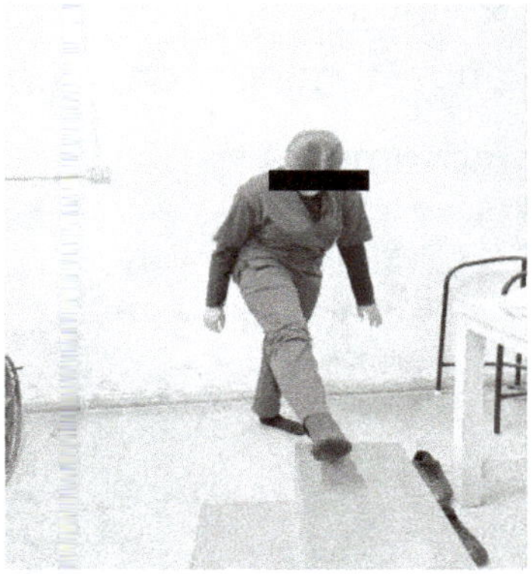	10.	**Leg movements**- In supine position, lift one leg then both leg lifting and alternate leg lifting. 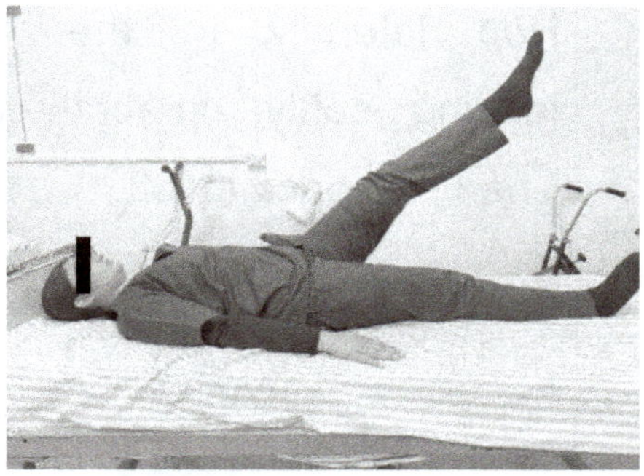
11.	**Abdomen stretch**- In prone lying position lifting both the legs at 45 degrees and lift your head 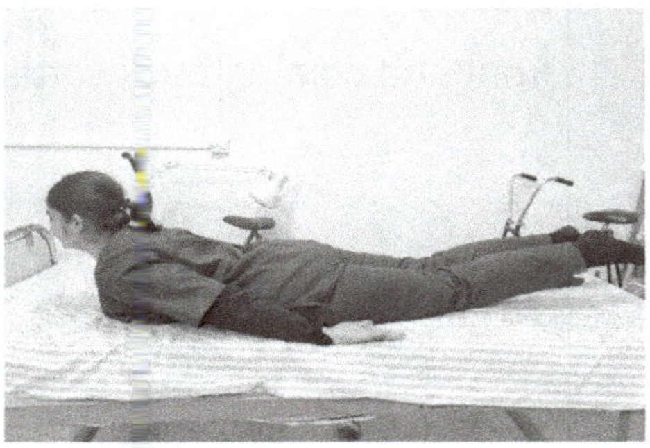	12.	**Back stretches**- In supine position, flex both the lower limbs towards the abdomen, and bring your head closer 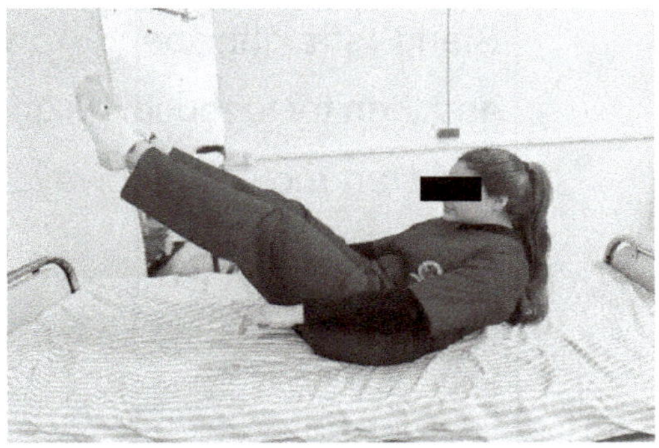

Continued...

3.14. Aerobic Exercises

Table 3.14.1. Details protocol of aerobic exercises

Sr. No	Body Movement	Sr. No	Body Movement
1.	Stand with feet hip width apart. Hop laterally to the right, landing softly on both feet. Then hop back to left. 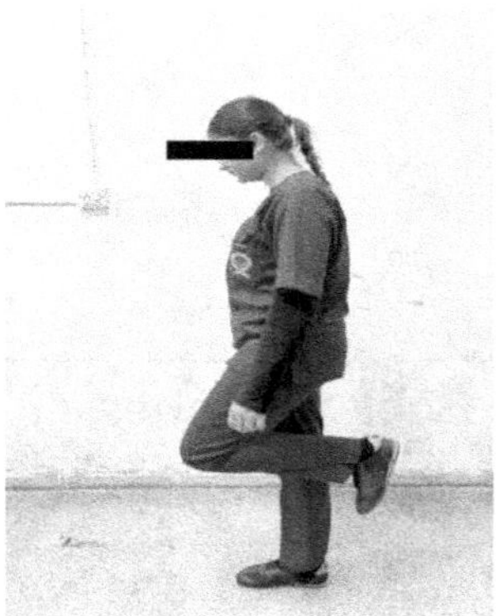	2.	Sitting position, flex both the knees and turning on the sides-fast movements 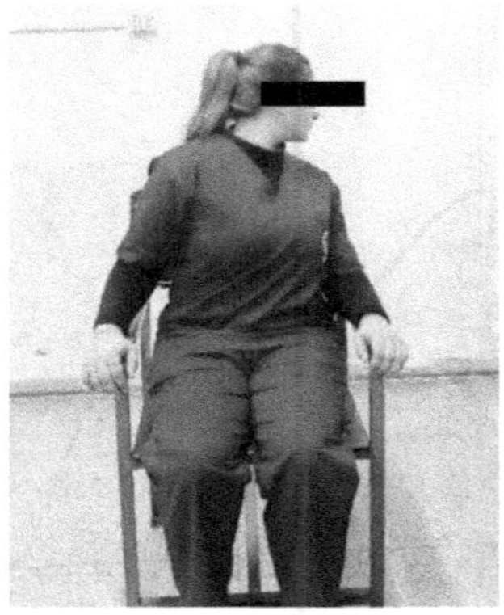
3.	Stand straight, bend forward and then try to touch the ground and start moving sideways 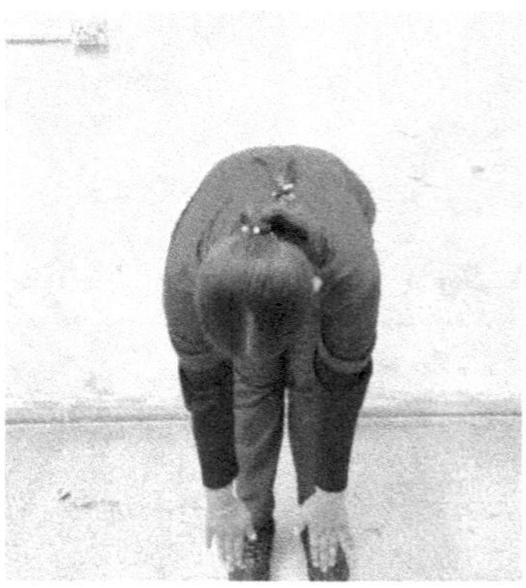	4.	Moving forward by kicking the leg front and coming backwards 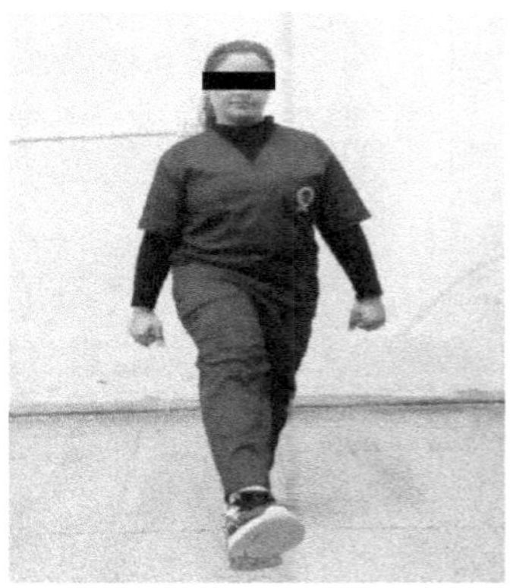

Continued...

5.	Moving forward while jogging 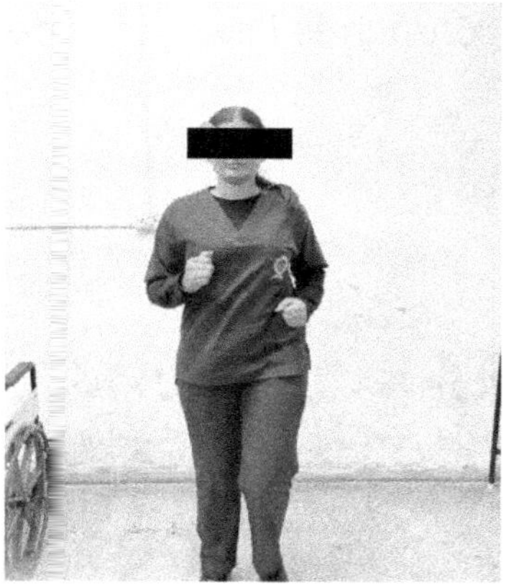	6.	Jogging and kicking forward; move four steps in form of letter v shape 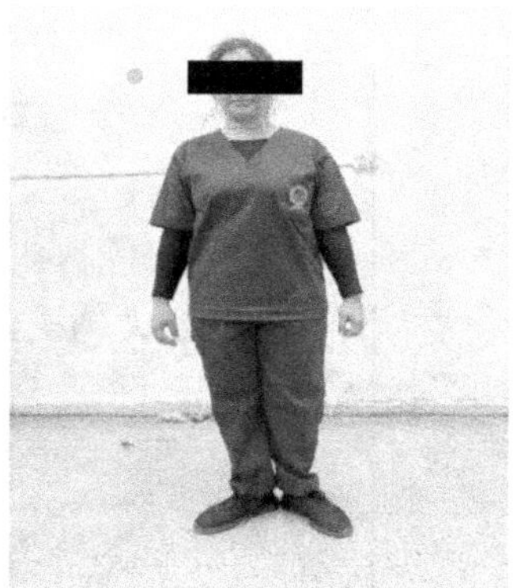
7.	Put right hand on opposite shoulder slight stretch. Repeat same with left hand 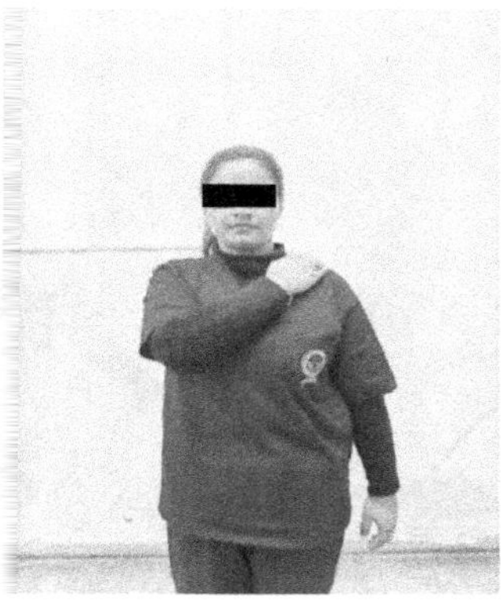	8.	In prone position; extension of spine 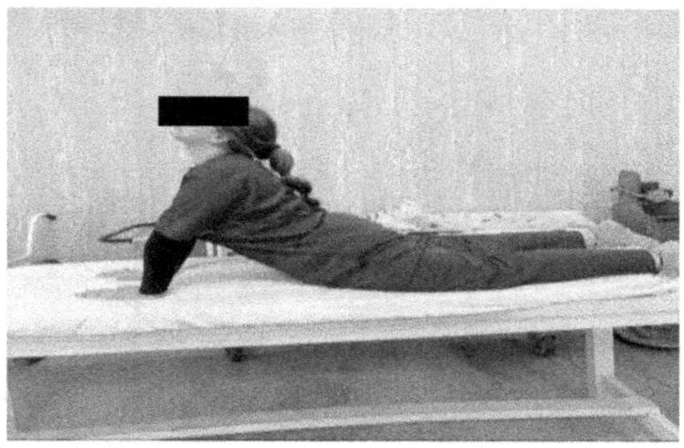
9.	Bend towards right with abdominal crunches 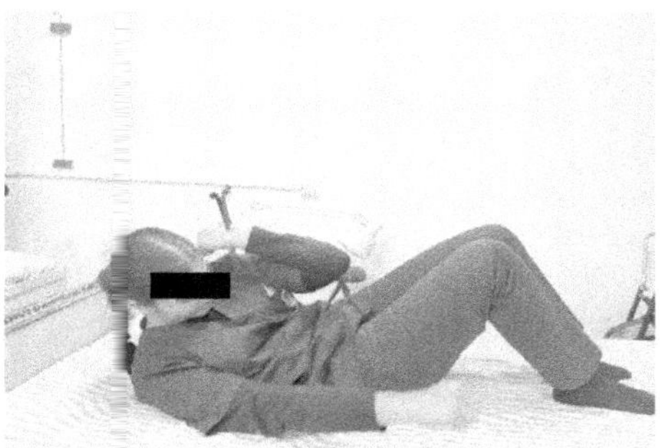	10.	In supine position; abdominal crunches bend right side 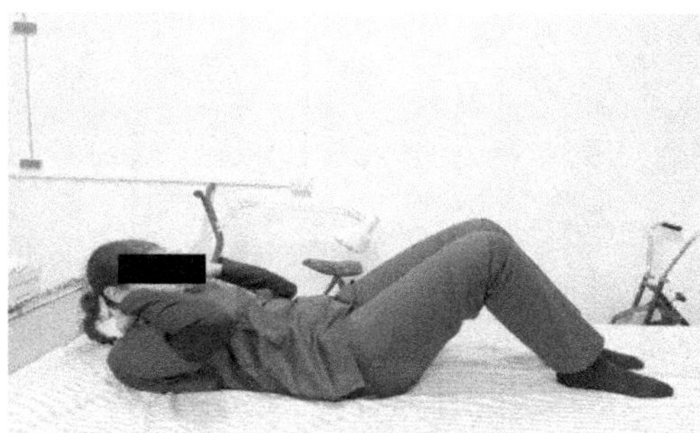

Continued...

11.	Hop up towards left 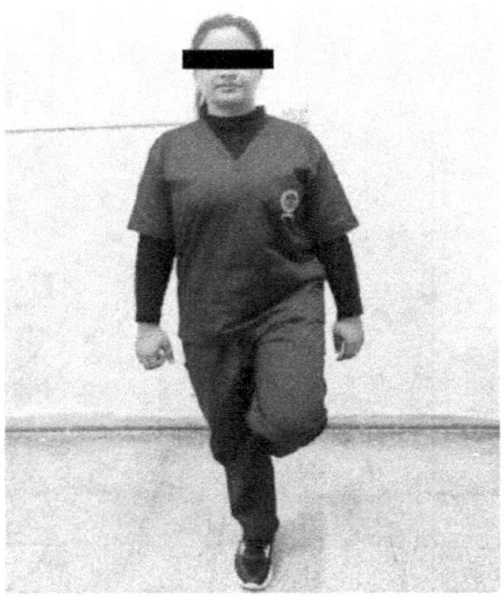	12.	In supine position; abdominal crunches bend left side 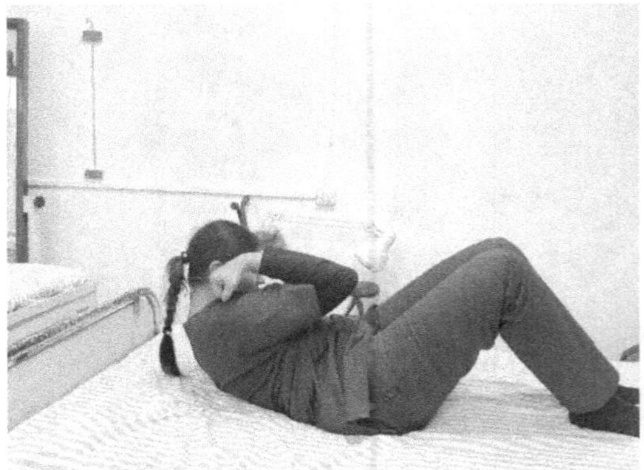
13.	In standing, swinging hands and leg movements (Marching) 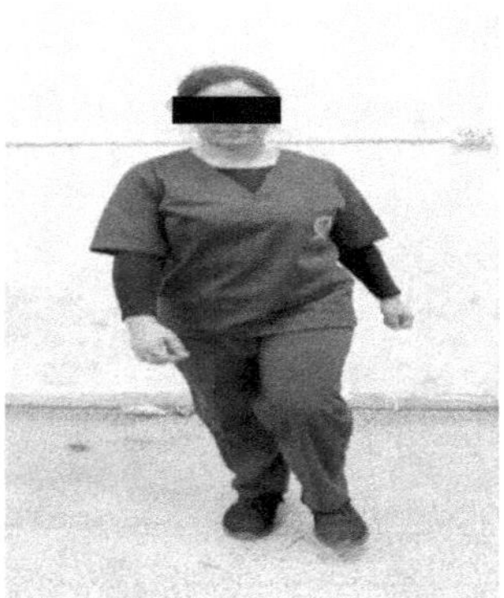	14.	Bend forward and swinging hands front and backward. 

Continued...

15.	Sitting position turning laterally on both sides alternatively-slow movements 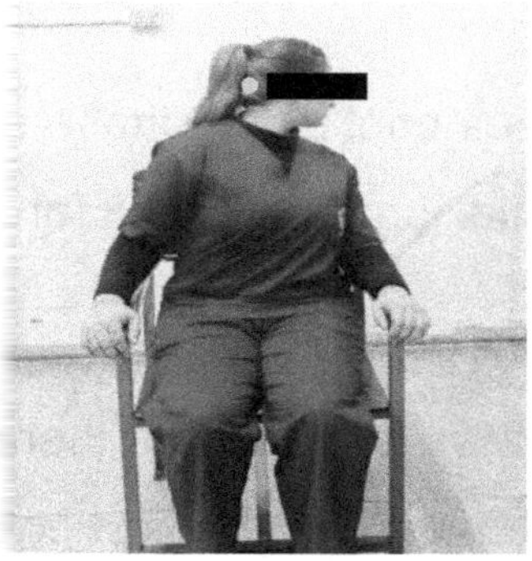	16.	Sitting position turning laterally on both sides alternatively-fast movement 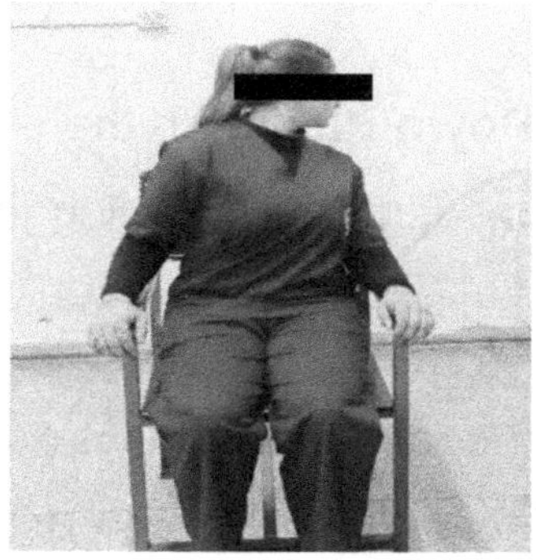
17.	Upward reaches; open and closing the hands and flex and extend elbow 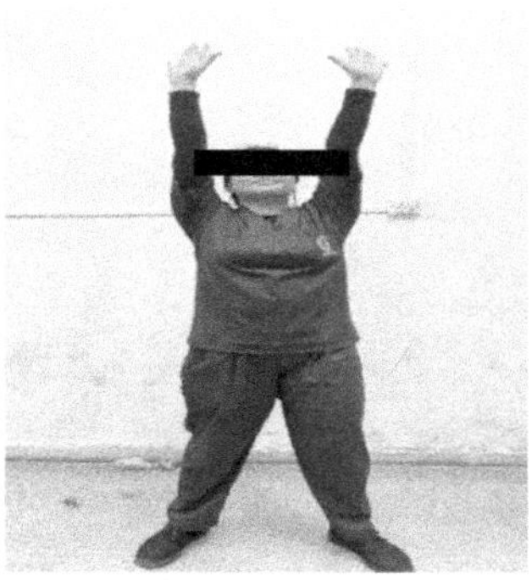	18.	Upward reaches with extend elbow; move wrist in clockwise and anti-clockwise 
19.	Hip in slight flexion; take feet in front; move ankle clockwise and anticlockwise 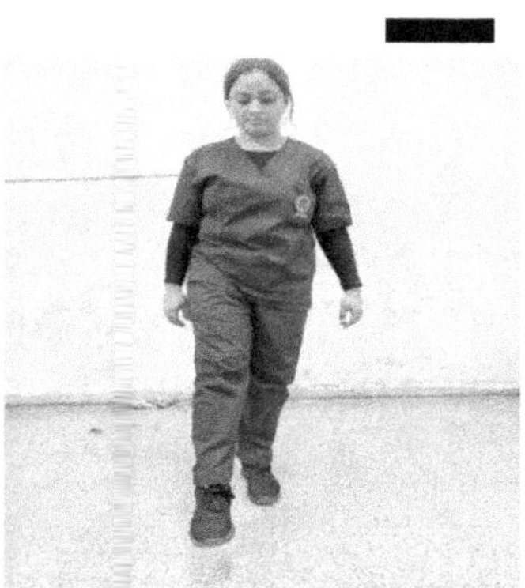	20.	Cross sitting position, flexing both hip and knee and turning on the sides-slow movements 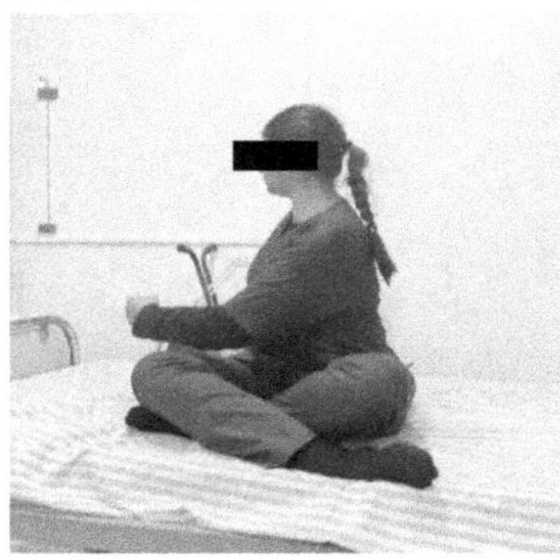

The participants were advised not to do any exercises during menstruation period. Instructions were given to drink lots of water, perform every movement 16 times. Two sets of above exercises were performed for 30 minutes.

Cool Down Phase: In the end of aerobic exercises cool down phase started with slow ankle movements, slow jogging at one place and flexion of the trunk forward. Cool down phase was for 10 minutes.

On the completion of last day of session of cognitive behavioural therapy and aerobic exercises a lecture on menstrual hygiene was given and filled following tools:

- Pain Disability Index Questionnaire.
- Health related short form-36 quality of life questionnaire.
- Beck Depression Inventory.
- Beck Anxiety Inventory.

3.13. Follow Up

All the participants were asked to start maintaining DRSP dairy for only luteal phase for next 6 months. Every month the investigator collected the record.

A summary of the methodology followed for completing the research is given in figure 3.6 as a flow chart.

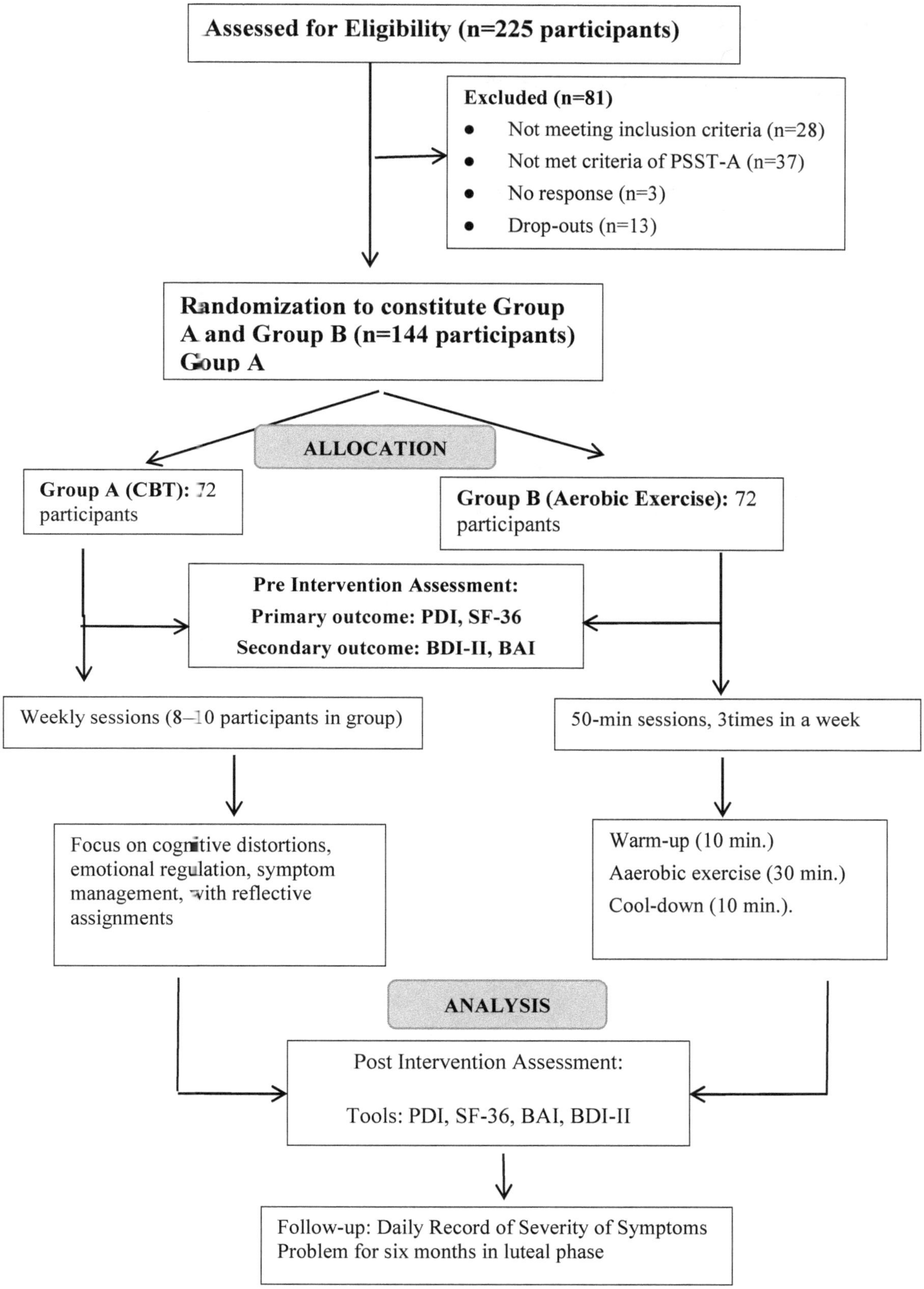

3.5. Flow chart of methodology followed

3.14. Statistical analysis

The collected data was calculated by student's t test.

$$f(t) = \frac{\Gamma\left(\frac{\nu+1}{2}\right)}{\sqrt{\nu\pi}\,\Gamma\left(\frac{\nu}{2}\right)}\left(1+\frac{t^2}{\nu}\right)^{-\frac{\nu+1}{2}}$$

The effect size was calculated with Cohen's d formula

$$d = \frac{M_2 - M_1}{\sqrt{\frac{SD_1{}^2 + SD_2{}^2}{2}}}$$

4.1. Sociodemographic characteristics of study participants

The present study was carried out on resident girl students of constituent Colleges of Adesh University located at village Bhuccho Kalan, District Bathinda, Punjab, during academic session 2022-2023. The aim of this study was to assess the impact of cognitive behavioral therapy (CBT) and aerobic exercise on premenstrual syndrome (PMS) in adolescents. The 144 participants finally selected for study, were assigned to two groups. Group A, received group-based CBT for eight weeks (CBT Group), and Group B, participated in aerobic exercise three times per week (Aerobic exercise Group). Observations were recorded for primary outcomes for CBT and aerobic exercises. Pain Disability Index and the domains of RAND Sf-36 Health-Related Quality of Life Questionnaire, as well as secondary outcomes, such as the Beck Anxiety Inventory and the Beck Depression Inventory-II, before and after the intervention. Effectiveness of both interventions was examined by within-group comparisons using the paired Student's t-test, while between-group comparisons were analyzed using the unpaired Student's t-test.

Upon initial contact with students, the sociodemographic data sheet was given to all enrolled participants. The sociodemographic data sheet consisted of information on current age, menarche age, parental income and education, permanent address, menstrual characteristics like menstrual cycle duration, perceived menstrual flow, frequency of changing absorbent, absorbent type, menstrual hygiene, cleaning practices during menstruation and any pre-diagnosed underlying disease as per (Annexure-I). The height and weight of each participant were measured to calculate body mass index (BMI) and recorded in the same sheet. All the participants were explained the terms used in the demographic data sheet and they were asked to fill this sheet in a face to face session.

Out of 225 participants initially approached for the study, 23 were excluded for not residing in the hostel. This left 202 participants, 05 were further excluded due to irregular menstruation. Among the remaining 197 eligible participants, 03 were excluded for not completing the daily severity of symptoms diary, the remaining 194 participants completed the PSST-A questionnaire, but 37 did not meet the criteria and were excluded, leaving 157 participants (81.11%) diagnosed with premenstrual syndrome. Further, these candidates were given PMSS form to assess the impact of PMS on them. These 157 candidates, 154 candidates had mild impact of PMS and three had moderate impact of PMS. Thirteen participants (8.3%) dropped out of the study for various reasons. The remaining 144 participants finally completed the study. The flow chart explaining the details on recruitment of the participants for the present study is given in figure 4.1.

Number of students contacted (225)

↓

Day scholars excluded (23)

↓

Excluded due to irregular menstrual cycle (05)

↓

Daily Severity Symptoms Diary (DRSP) distributed to 197

↓

Excluded for not filling Daily Severity Symptoms Diary (DRSP) (03)

↓

Number of participants filling PSST-A (194)

↓

Excluded as per guideline of ACOG and DSM-V criteria (37)

↓

Number of participants diagnosed with PMS (157)

↓

Number of participants dropped off (13)

↓

Number of students finally recruited as participants (144)

↓

Constitution of groups

↙ ↘

Group CBT (72) | Group Aerobic Exercises (72)

Figure 4.1. Details of methodology used for recruiting of participants for study

4.1.1 Prevalence of Premenstrual Syndrome

The recorded observations revealed that 194 participants (86.2%) who completed the DRSP diary for two months and met the inclusion criteria. Among them, 37 participants (19.0%) did not meet the PSST-A score criteria, indicating the absence of PMS. Consequently, 157 participants were diagnosed with premenstrual syndrome based on the PSST-A questionnaire scores, resulting prevalence rate of PMS was 81.0% in the given study population. The participants of the study exhibited following premenstrual syndrome symptoms.

Table 4.1.1. List of common PMS symptoms reported by the participants

Sr. No.	Common symptoms	Sr. No.	Common symptoms
1.	Anger	10.	Hypersomnia
2.	Anxiety	11.	Somnia
3.	Depressed mood	12.	Feeling of overwhelmed
4.	Fatigue	13.	Difficulty in concentrating
5.	Weight gain	14.	Over eating
6.	Breast tenderness	15.	Decreased interest in work
7.	Headache	16.	Decreased interest in domestic chose
8.	Joint pain	17.	Decreased interest in social activities

9.	Tearfulness		

The premenstrual symptoms scale scores of 157 participants revealed that 154 (99%) participants experienced mild impact of PMS symptoms while only 03 (1%) participants reported moderate impact on them. None of the participants exhibited severe impact of PMS symptoms. The following symptoms were reported by the participants as per the severity of impact.

Table 4.1.2. List of symptoms according to impact of severity reported by participants

Sr. No.	**Severe symptoms**	**Sr. No.**	**Moderate symptoms**	**Sr. No.**	**Mild symptoms**
1.	Anger	1.	Anxiety	1.	Feeling overwhelmed
2.	Tearfulness	2.	Hypersomnia	2.	Decreased interest in social activities
3.	Depressed mood	3.	Decreased interest in work	3.	Decreased concentration
4.	Fatigue	4.	Decreased interest in domestic chose		
5.	Physical symptoms				
6.	Weight gain				
7.	Over eating				

8.	Insomnia				

4.1.3. Age of attainment of menarche

The age of initiation of menarche varied from 10 years to 17 years among 144 participants included in the study. Early menarche was seen at the age of 10 years and 11 years, 05 (3.47%) and 02 (1.39%) respectively. 23 (15.97%) participants, had their first period at the age of 12 years, 53 (36.81%) participants had at the age of 13 years. 45 (31.25%) of participants experienced menarche at the age of 14 years. 12 (8.33 %,) participants experienced late menarche at the age of 15 years. Menarche was observed at the age of 16 and 17 years respectively, in 03 (2.08 %) and 01 (0.69%) participant. The variation in age of attainment of the menarche is shown in the figure 4.1.3.

4.1.4. Body mass index of the participants

Based on the BMI classification of the WHO, the participants were grouped into four categories; underweight (<18.5), over weight (18.25-24.9), normal weight (25-29.9) and obese (30 and above). Ninety participants (62.50%) had normal BMI, 25 participants (17.36%) were, underweight, 18 (12.50%) participants were classified overweight and 11 (7.64%) were categorized obese. The distribution of 144 participants in these four categories is shown in figure 4.1.4.

4.1.5. Span of menstruation cycle

The menstrual cycle of 144 participants varied from 24 days to 31 days (figure 4.1.5.). Out of these 116 (80.56%) participants had 30 days spread, 12 (8.33%) had 28 days' span, 07(4.86%) participants had 25 days' span and 03 (2.08%) participants had 24 days' span. The least common menstrual cycle observed was in 0 (0.69%) participant with 26 days of menstruation cycle. Few participants 02

(1.39%), 02 (1.39%) and 01(0.69%) participants had menstruation cycle span of 27, 29 and 31 days, respectively. The distribution of participants according to the span of menstruation cycle is shown in figure 4.1.5.

4.1.6. Perceived menstruation flow

The distribution of participants in moderate, heavy and scarcer flow categories are shown in figure 4.1.6. The 110 (78.47%) participants perceived that nature of menstruation flow was moderate (normal). Heavy flow was perceived by 32 (18.75%) participants. Whereas, only 02 (1.39%) participants felt that they have scarce menstruation flow during the study period.

4.1.7. Habit of changing absorbent

The behavior of participants regarding duration of change of absorbent during menstruation is shown in Figure 4.1.7. The 105 (73%) participants changed their absorbent every six hourly. 37 (26%) participants reported a 4-hourly change, while only 02 (01%) reported a 10-hourly change of absorbent.

4.1.8. Education status of the parents of the participants

The distribution of participants according to education of their parents is shown in Figure 4.1.8. The parents of all the participants were educated, 66 parents (45.83%) were having higher education and 78 (54.17%) were educated up to elementary level.

4.1.9. Employment status of the parents of the participants

The distribution of the participants according to the type of employment of their parents is shown in figure 4.1.10. Parents of the participants were working in government sector, private sector and self-employed. The parents of 76 participants (52.78%) were self-employed, the parents of 48 participants (33.33%)

were working in government sector and the parents of 20 participants (13.89%) were working in private sector.

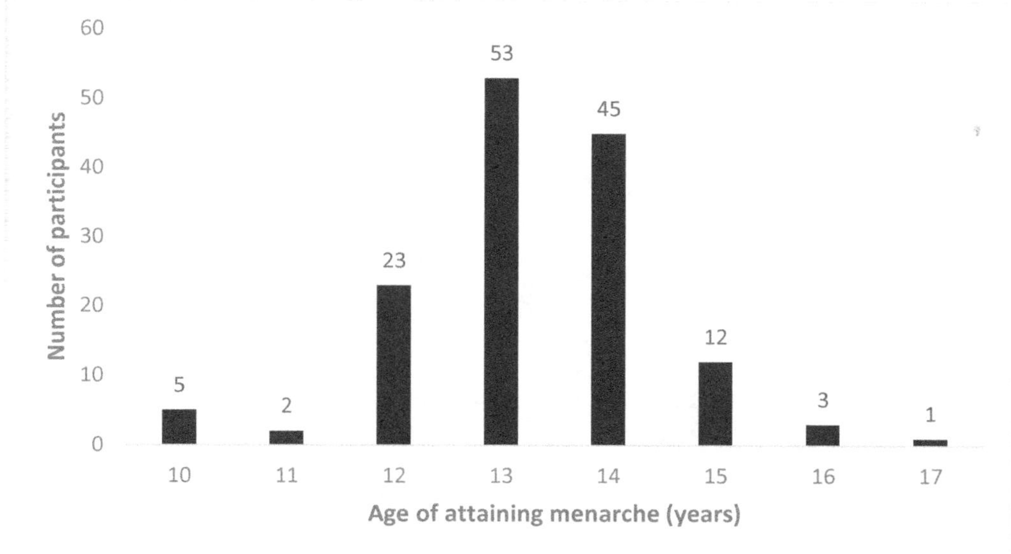

Figure 4.1.3. Distribution of participants according to age of attainment of menarche.

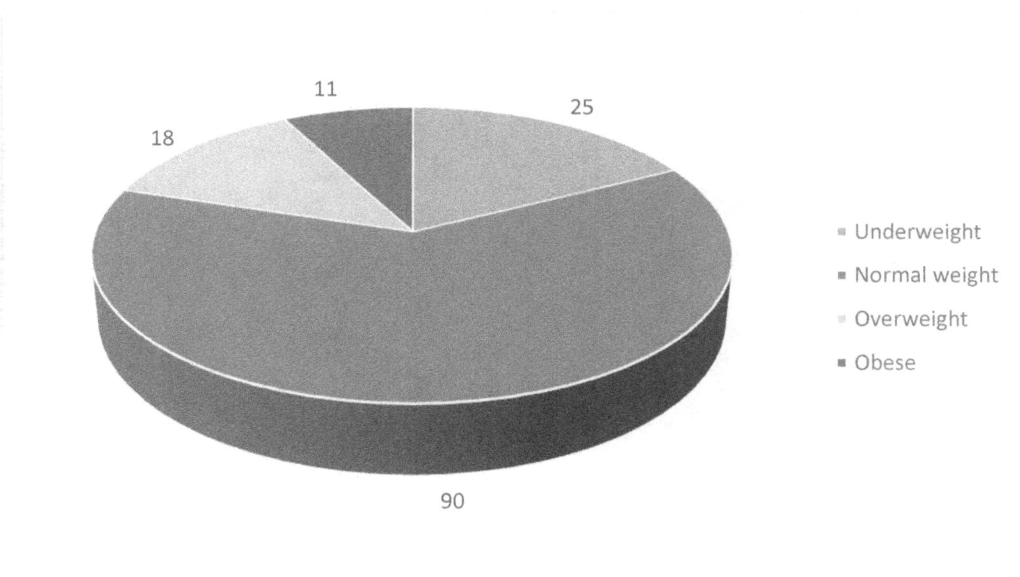

Figure 4.1.4. Distribution of participants according to body mass index

Figure 4.1.5. Distribution of participants according to span of menstruation cycle

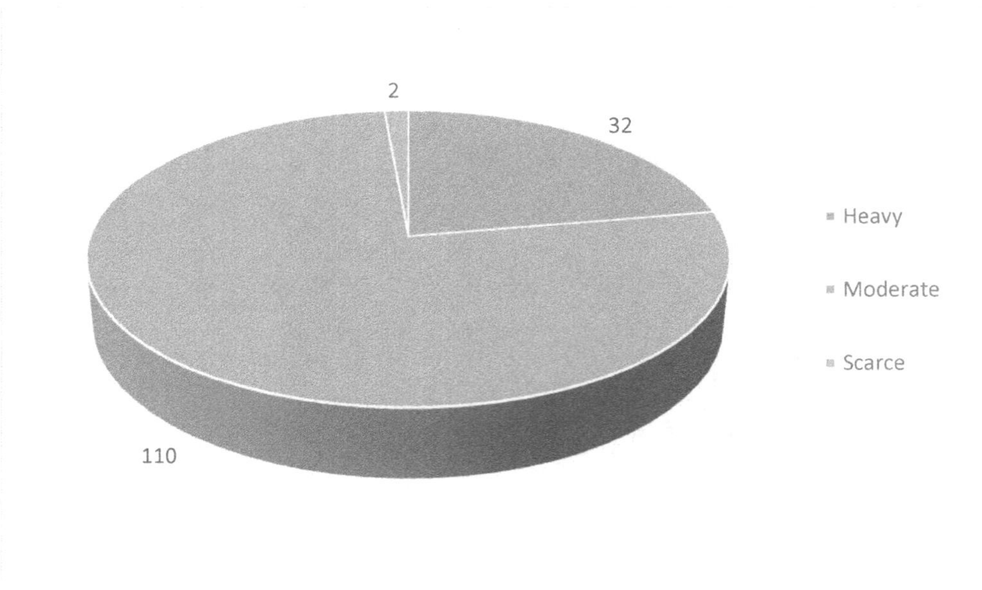

Figure 4.1.6. Distribution of participants according to perceived menstruation flow.

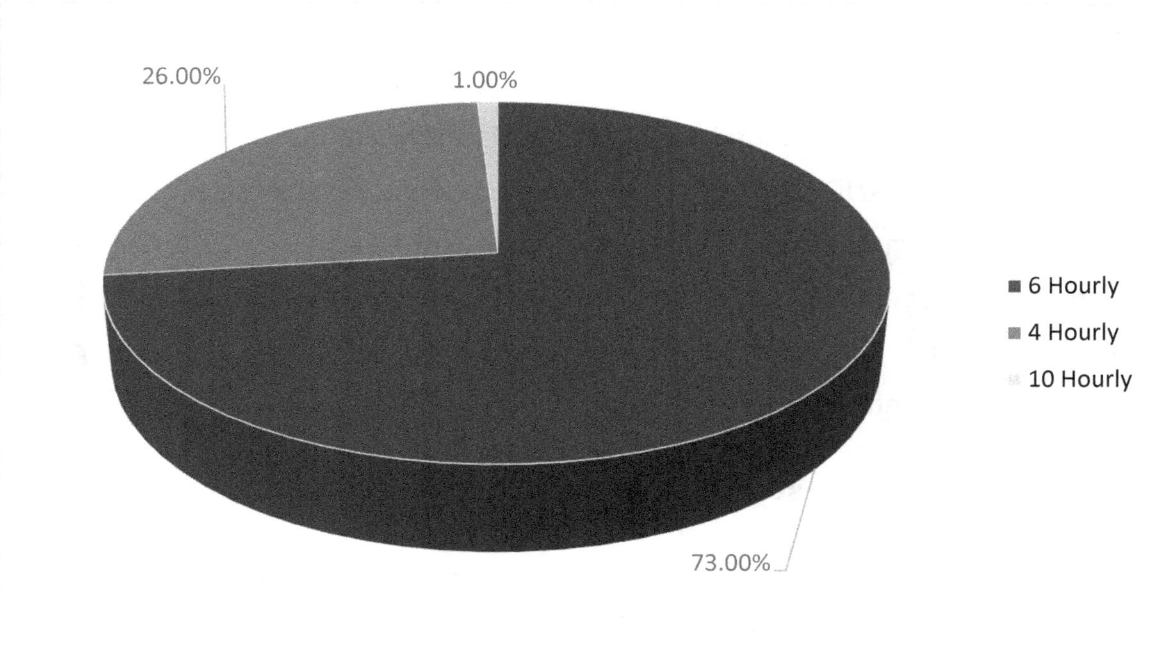

Figure 4.1.7. Distribution of participants according to habit of changing of absorbent

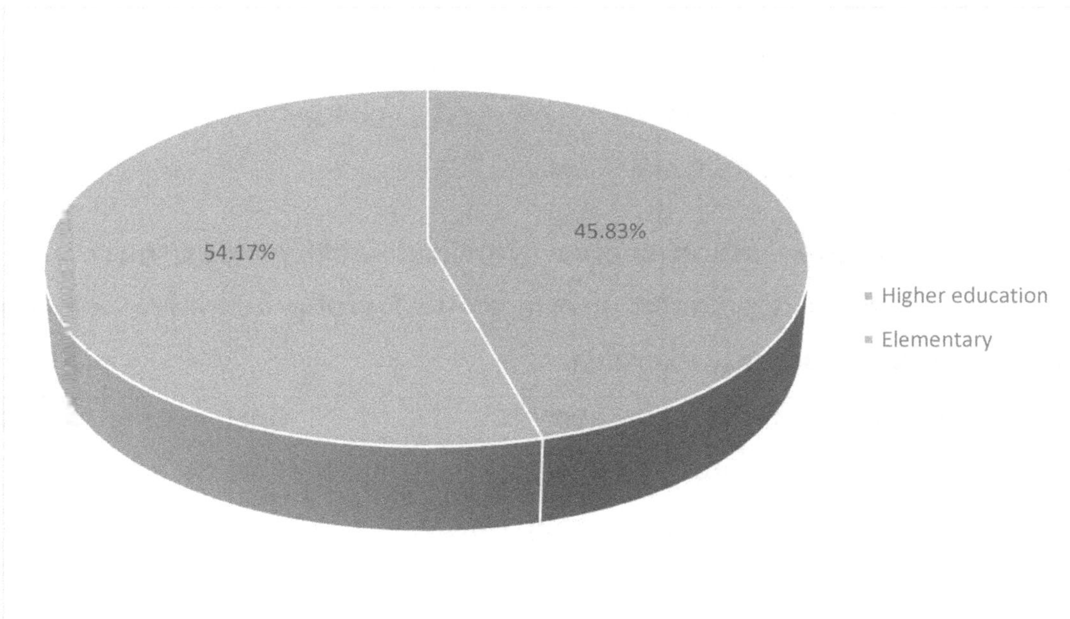

Figure 4.1.8. Distribution of the participants according to the education status of parents

4.2. Outcome measures for CBT and aerobic exercises

Participants of both groups who completed the daily severity problem diary were assessed for outcome measures before and after the interventions. The primary outcome measures included the Pain Disability Index and quality-of-life questionnaire, a RAND Short Form-36. Anxiety and depression were evaluated as secondary outcomes using the Beck Anxiety Inventory and the Beck Depression Inventory-II, respectively. After completing the 8-week interventions (CBT and aerobic exercise), the same outcome measures were assessed to determine the outcome of both interventions on PMS.

4.2.1. Pre-intervention scores

The pre-intervention scores for pain disability index, beck anxiety inventory and beck depression inventory-II of the CBT and aerobic exercises groups are given in table 4.2.1 and shown in figure 4.2.1. The mean pain disability index for CBT group sand aerobic exercise group was 51.39 and 53.62, respectively. The mean beck anxiety inventory score for CBT and aerobic exercise groups was 30.67 and 30.84, respectively. These scores for the two groups were not statistically different. Similarly, the beck depression inventory –II for CBT and aerobic exercise groups were 35.99 and 35.33, respectively. The scores for the two groups before the intervention were not different statistically.

Table 4.2.1. Pre-intervention scores of pain disability index, beck anxiety inventory and beck depression inventory-II of group cognitive behavioral therapy and group aerobic exercise.

Outcome measure	CBT group (Mean ± SD)	aerobic exercise group (Mean ± SD)	t-value	p-value	Conclusion

Pain Disability Index (PDI)	51.39 ± 20.02	53.62 ± 12.48	-0.47	0.64	Non-significant
Beck Anxiety Inventory (BAI)	30.67 ± 13.84	30.84 ± 13.83	-0.07	0.90	Non-significant
Beck Depression Inventory-II (BDI-II)	35.99 ± 10.83	35.33 ± 12.08	-0.10	0.92	Non-significant

The pre-intervention means score of physical function domain of Sf (36), for both the CBT and aerobic exercise groups were 46.53 and 48.43, respectively which were not statistically significant. The mean score of role limitation due to physical health domain were 28.13 and 28.82 for CBT and aerobic exercise groups respectively, were statistically not significant. The means scores of role limitation due to emotional problems domain were 26.22 and 24.09 respectively, for CBT and aerobic exercise groups and these were also not statistically significant. Similarly, the mean scores of domains like energy/fatigue (45.69 and 44.72), emotional well-being (47.28 and 49.51), social functioning (43.34 and 46.46), pain (37.10 and 38.39), general health (47.50 and 50.00) and health change (39.86 and 44.97), for CBT and aerobic exercise groups were not statistically different, respectively. The observation given in table 4.2.2 and shown in figure 4.2.2 suggest that the groups following CBT and Aerobic exercise of 72 participants each are comparable and possess same baseline health status before the interventions.

Table 4.2.2. Pre-intervention scores for short form 36 (Sf36) domains of cognitive behavioral therapy and aerobic exercise groups.

Sf (36) Domains	CBT group Mean ±SD	Aerobic exercise group Mean± SD	t-value	p-value	Conclusion
Physical Function	46.53 ±22.29	48.43± 21.87	-0.51	0.61	Non-significant
Role Limitation Due to Physical Health	28.13 ±21.97	28.82 ±28.66	-0.16	0.87	Non-significant
Role Limitation Due to Emotional Problems	26.22 ±24.04	24.09 ±24.74	0.52	0.60	Non-significant
Energy / Fatigue	45.69 ±11.51	44.72 ±11.53	0.50	0.61	Non-significant
Emotional Well-Being	47.28 ±11.36	49.51 ±12.50	-1.12	0.26	Non-significant

Social Functioning	43.34 ±12.22	46.46 ±12.30	-1.52	0.13	Non-significant
Pain	37.10 ±10.11	38.39 ±18.68	-0.51	0.61	Non-significant
General Health	47.50 ±13.10	50.00 ±16.61	-1.00	0.32	Non-significant
Health Change	39.86 ±19.76	44.97 ±19.04	-1.58	0.12	Non-significant

4.2.2 Post-interventional scores

The comparison of pain disability index (PDI), beck anxiety inventory (BAI) and beck depression inventory-II (BDI-II) scores of participants are given in table 4.2.2.1 and shown in figure 4.2.2.1.

Pain Disability Index (PDI) - The mean score of PDI for CBT group was 51.39 prior to intervention which was reduced to 26.92 after CBT.

Pain Disability Index (PDI) - The mean score of PDI for CBT group was 51.39 prior to intervention which was reduced to 26.92 after CBT.

Beck Anxiety Inventory (BAI) - The mean scores of BAI for CBT group was 30.67 prior to the intervention which was reduced to 15.48 post intervention.

Beck Depression Inventory-II (BDI-II) - The mean scores of BDI-II for CBT group prior to intervention was 34.99 and post intervention has been reduced to 19.03.

The reduction in the scores of all the three outcomes after intervention is significant suggesting a positive effect of CBT in reducing the pain disability, anxiety and depression due to PMS in the participants of the study.

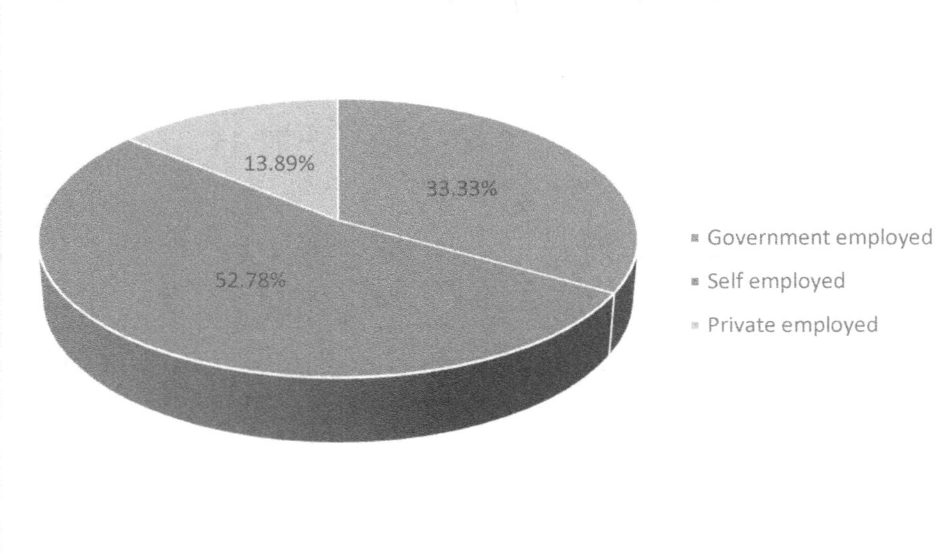

Figure 4.1.9. Distribution of the participants according to employment of their parents.

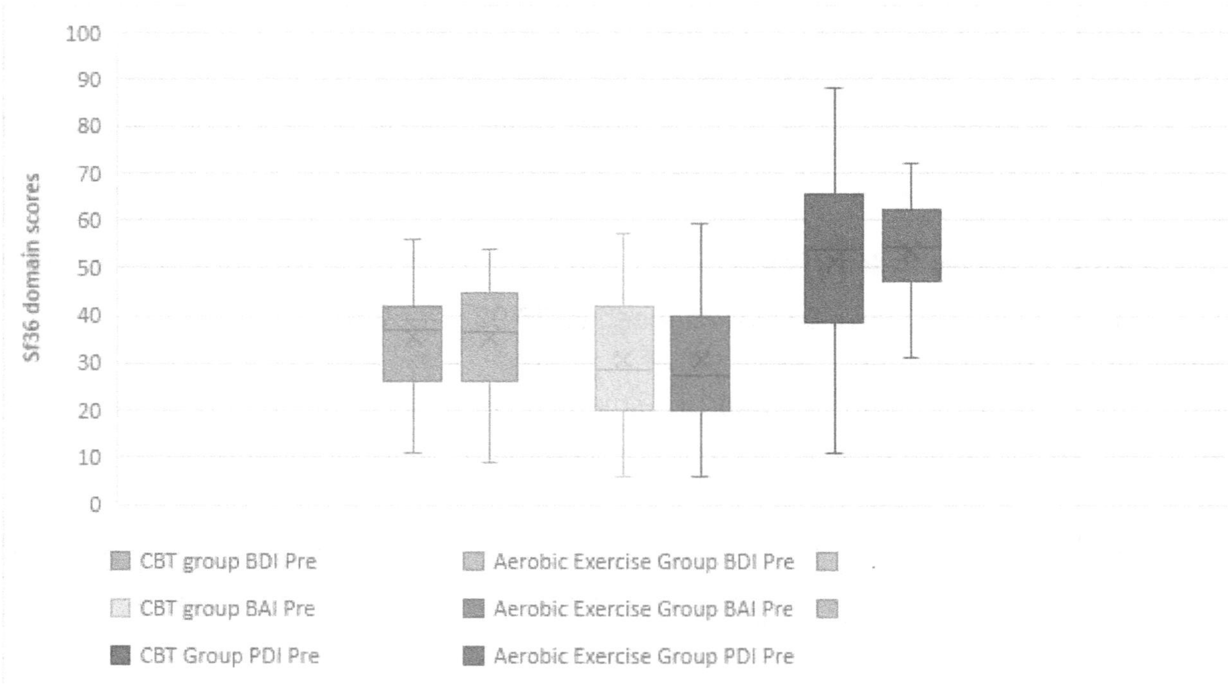

Figure 4.2.1. Pre-intervention scores of pain disability index, beck anxiety inventory and beck depression inventory-II of group cognitive behavioral therapy and group aerobic exercise.

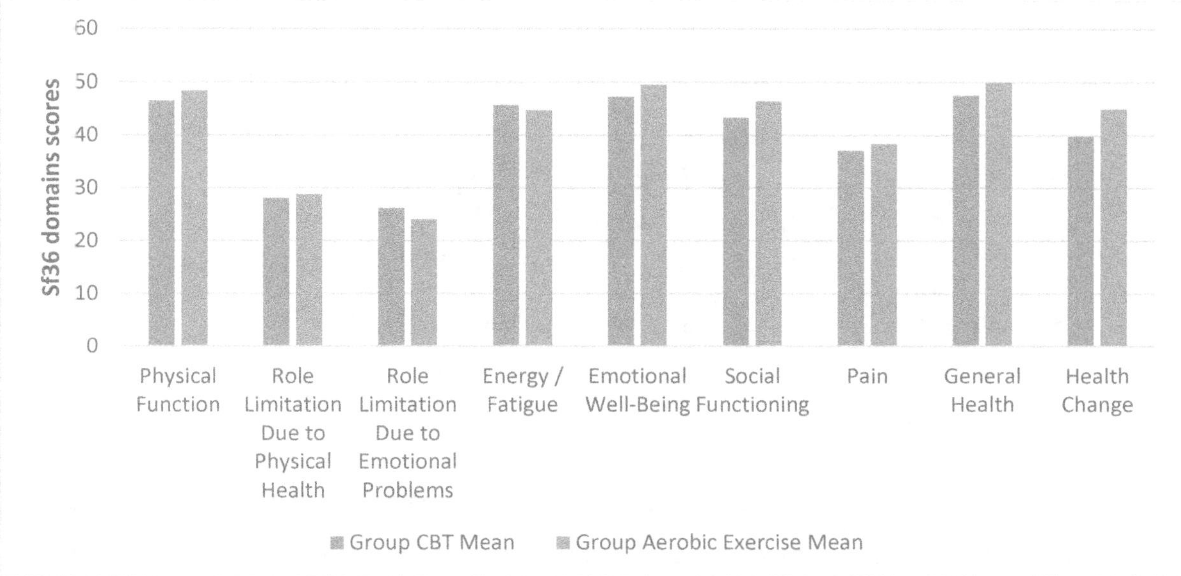

Figure 4.2.2. Pre-intervention scores for short form 36 (Sf36) domains of cognitive behavioral therapy and aerobic exercise group.

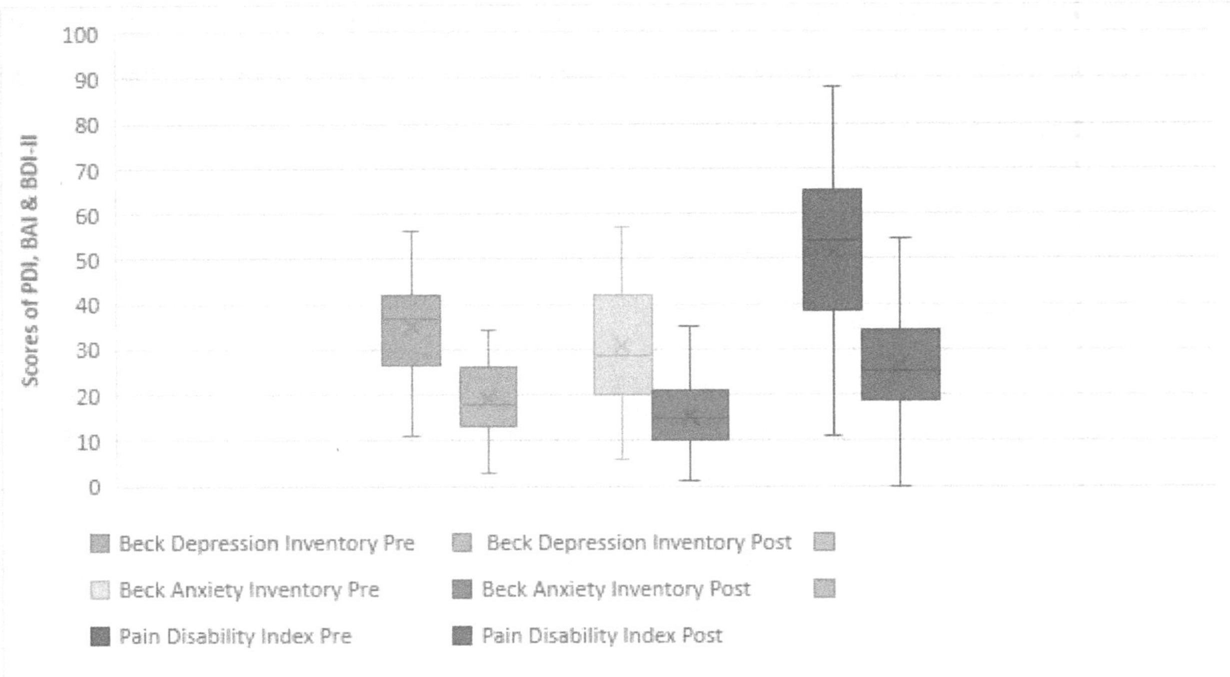

Figure 4.2.2.1. Pre and post-intervention scores of pain disability index, beck anxiety inventory and beck depression inventory-II of cognitive behavioral therapy (CBT) group

Table 4.2.2.1. Pre and post-intervention scores of pain disability index, beck anxiety inventory and beck depression inventory-II of cognitive behavioral therapy (CBT) group.

Outcome measures	CBT group (Mean ± SD) Pre-intervention	CBT group (Mean ± SD) Post - intervention	t-value	p-value	Conclusion
Pain Disability Index (PDI)	51.39 ±20.02	26.92 ± 12.19	8.09	<0.001	Highly significant
Beck Anxiety Inventory (BAI)	30.67 ±13.84	15.48 ±7.93	7.38	<0.001	Highly significant
Beck Depression Inventory-II (BDI-II)	34.99 ±10.83	19.03 ± 8.58	8.09	<0.001	Highly significant

The mean scores of nine domains of Sf36 pre and post intervention are given in table 4.2.2.2 and shown in the figure 4.2.2.2. The mean score of physical function domain of Sf36 was 46.53 prior to the intervention. The post-intervention mean score was 75.14. The mean score of Sf36, role limitation due to physical health domain was 28.13 prior to the intervention and mean score post intervention has been improved to 84.51. The means score of role limitation due to emotional problems domain was 26.22 prior to the intervention and post intervention mean scores for this domain improved to 77.06. The mean score of energy/fatigue domain was 45.69 prior to the intervention and post intervention mean score for the domain has been improved to 50.06. The mean score of emotional well-being domain prior to the intervention was 47.28 and post intervention mean scores increased to 60.59. The mean score of social functioning domain prior to the intervention was 43.34 and post intervention mean score improved to 64.43. The mean score of pain domain prior to the intervention was 37.10 and post intervention mean score improved to 60.59. The mean score\ of general health domain prior to intervention was 47.50 and post intervention mean score has been improved to 57.50. Similarly, the mean score of health change domain prior to intervention was 39.86 and post intervention mean sores improved to 64.6. The mean scores of all the Sf36 domains improved due to intervention and they were statistically significant.

Table 4.2.2.2. Pre-intervention and post-intervention scores for Short form 36 (Sf36) domains of cognitive behavioral therapy (CBT) group.

Sf (36) Domains	**CBT group (Mean ± SD) Pre-intervention**	**CBT group (Mean ± SD) Post - intervention**	**t-value**	**p-value**	**Conclusion**
Physical Function	46.53 ± 22.29	75.14 ± 19.73	-11.38	<0.001	Differ significantly

Role Limitation Due to Physical Health	28.13 ± 21.97	84.51 ± 20.10	-15.69	<0.001	Differ significantly
Role Limitation Due to Emotional Problems	26.22 ± 24.04	77.06 ± 31.46	-10.44	<0.001	Differ significantly
Energy / Fatigue	45.69 ± 11.51	59.06 ± 11.18	-7.72	<0.001	Differ significantly
Emotional Well-Being	47.28 ± 11.36	60.47 ± 12.44	-7.72	<0.001	Differ significantly
Social Functioning	43.34 ±12.22	64.43 ± 14.44	-9.39	<0.001	Differ significantly
Pain	37.10 ±10.11	60.59 ± 17.56	-11.52	<0.001	Differ significantly
General Health	47.50 ±13.10	57.50 ± 10.90	-6.77	<0.001	Differ significantly
Health Change	39.86 ±19.76	64.69 ± 17.61	-7.58	<0.001	Differ significantly

The effect of aerobic exercises on reducing pain disability index (PDI), beck anxiety inventory (BAI) and beck depression inventory-II (BDI-II) mean scores of participants are given in table 4.2.2.3 and shown in figure 4.2.2.3.

The mean score of PDI of the participants before aerobic exercises for group was 52.72 and which was reduced to 28.95 after aerobic exercises. This parameter has shown significant reduction in pain disability index. Similarly, the mean score of BAI and BDI-II of the participants before aerobic exercise was 30.85 and 34.99 which reduced to 17.61 and 19.03 respectively. These results suggest that aerobic exercises cause significant reduction in the pain disability, anxiety and depression observed due to PMS.

Table 4.2.2.3. Pre-intervention and post-intervention scores of pain disability index, beck anxiety inventory and beck depression inventory-II participants in aerobic exercise group.

Outcome measure	Aerobic exercise group (Mean ± SD) pre-intervention	Aerobic exercise group (Mean ± SD) post-intervention	t-value	p-value	Conclusion
Pain Disability Index (PDI)	52.72 ± 12.48	28.95 ± 13.06	13.44	<0.001	Differ significantly
Beck Anxiety Inventory (BAI)	30.85 ± 13.83	12.88 ± 5.76	12.82	<0.001	Differ significantly
Beck Depression Inventory-II (BDI-II)	35.33 ± 12.08	17.61 ± 8.45	12.15	<0.001	differ significantly

The impact of aerobic exercises on nine domains of Sf (36) is given in table 4.2.2.4 and shown in the figure 4.2.2.4. The mean scores for all the nine domains of Sf (36) improved significantly after exercises. Maximum improvement was seen with respect to domain (i) role limitation due to emotional problems with mean 24.09 changing to 82.93, respectively and domain (ii) role limitation due to physical health with mean score changing from 28.82 to 78.13, respectively.

Table 4.2.2.4. Pre-intervention and post-intervention scores of domains of Short form 36 (Sf36) for aerobic exercise group.

Sf (36) Domain	Aerobic exercise group (Mean ± SD) Pre-intervention	Aerobic exercise group (Mean ± SD) Post-intervention	t-value	p-value	Conclusion
Physical Function	48.43± 21.87	76.19 ± 15.75	-11.06	<0.001	Differ significantly
Role Limitation Due to Physical Health	28.82 ±28.66	78.13 ± 28.43	-10.28	<0.001	Differ significantly
Role Limitation Due to Emotional Problems	24.09 ±24.74	82.93 ± 26.27	-13.12	<0.001	Differ significantly
Energy / Fatigue	44.72 ±11.53	62.08 ± 10.86	-9.57	<0.001	Differ significantly
Emotional Well-Being	49.51 ±12.50	61.88 ± 12.08	-8.26	<0.001	Differ significantly
Social Functioning	46.46 ±12.30	68.02 ± 15.36	-9.28	<0.001	Differ significantly
Pain	38.39 ±18.68	64.65 ± 18.03	-8.82	<0.001	Differ significantly
General Health	50.00 ±16.61	61.39 ± 10.88	-5.65	<0.001	Differ significantly
Health Change	44.97 ±19.04	69.07 ± 20.82	-9.04	<0.001	Differ significantly

4.2.3. Comparative effect of CBT and aerobic exercise

The mean post intervention scores for pain disability index (PDI), beck anxiety inventory (BAI) and beck depression inventory-II (BDI-II) are given in table 4.2.3.1 and shown in figure 4.2.3.1. The outcome of both interventions, measure as pain disability index (PDI), beck anxiety inventory (BAI) and beck depression inventory-II (BDI-II) scores are not different statistically. Although the mean score of pain disability index for both CBT and aerobic exercise groups was low but post intervention of aerobic exercise pain disability reduced more as compare to CBT. Whereas, the scores of BAI and BDI-II of CBT group were better than aerobic exercise group, although the difference was non-significant.

Table 4.2.3.1. Between group comparisons of post- intervention mean scores of pain disability index, beck anxiety inventory and beck depression inventory-II of CBT and aerobic exercise groups.

Outcome measure	CBT group (Mean ± SD) Post-intervention	Aerobic exercise group (Mean ± SD) Post-intervention	t-value	p-value	Conclusion
Pain Disability Index (PDI)	26.92 ± 12.19	28.95 ± 13.06	-0.62	0.53	Not different
Beck Anxiety Inventory (BAI)	15.48 ±7.93	12.88 ± 5.76	1.45	0.15	Not different
Beck Depression Inventory-II (BDI-II)	19.03 ± 8.58	17.61 ± 8.45	0.65	0.52	Not different

The comparison of post-intervention scores of nine domains of Sf (36) is given in table 4.2.3.2 and shown in figure 4.2.3.2. The post intervention score of all the Sf (36) domains was higher for aerobic exercise group except for the "role limitation due to physical health" indicating that aerobic exercise is slightly better strategy for alleviating the effect of PMS as compared to CBT.

Table 4.2.3.2. Comparisons of post-intervention scores of nine domains of Sf (36) after CBT and aerobic exercise groups.

Sf (36) Domains	CBT group (Mean ± SD) post-intervention	Aerobic exercise group (Mean ± SD) post-intervention	t-value	p-value	Conclusion
Physical Function	75.14 ± 19.73	76.19 ± 15.75	-0.68	0.50	Interventions have same effect
Role Limitation Due to Physical Health	84.51 ± 20.10	78.13 ± 28.43	1.03	0.30	Interventions have same effect
Role Limitation Due to Emotional Problems	77.06 ± 31.46	82.93 ± 26.27	-1.26	0.21	Interventions have same effect
Energy / Fatigue	59.06 ± 11.18	62.08 ± 10.86	-1.46	0.14	Interventions have same effect
Emotional Well-Being	60.47 ± 12.44	61.88 ± 12.08	-0.79	0.43	Interventions have same effect
Social Functioning	64.43 ± 14.44	68.02 ± 15.36	-1.26	0.21	Interventions have same effect
Pain	60.59 ± 17.56	64.65 ± 18.03	-1.23	0.22	Interventions have same effect
General Health	57.50 ± 10.90	61.39 ± 10.88	-1.80	0.07	Interventions have same effect

Health Change	64.69 ± 17.61	69.07 ± 20.82	-1.13	0.26	Interventions have same effect

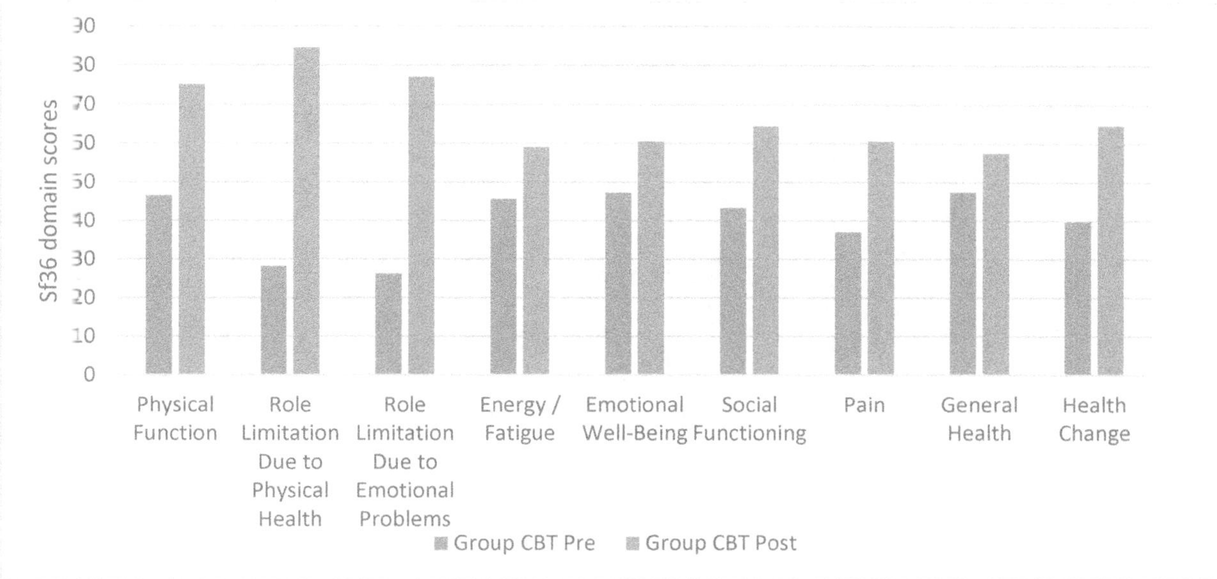

Figure 4.2.2.2. Pre-intervention and post-intervention scores of domains of Short form 36 (Sf36) for participants of cognitive behavioral therapy (CBT) group.

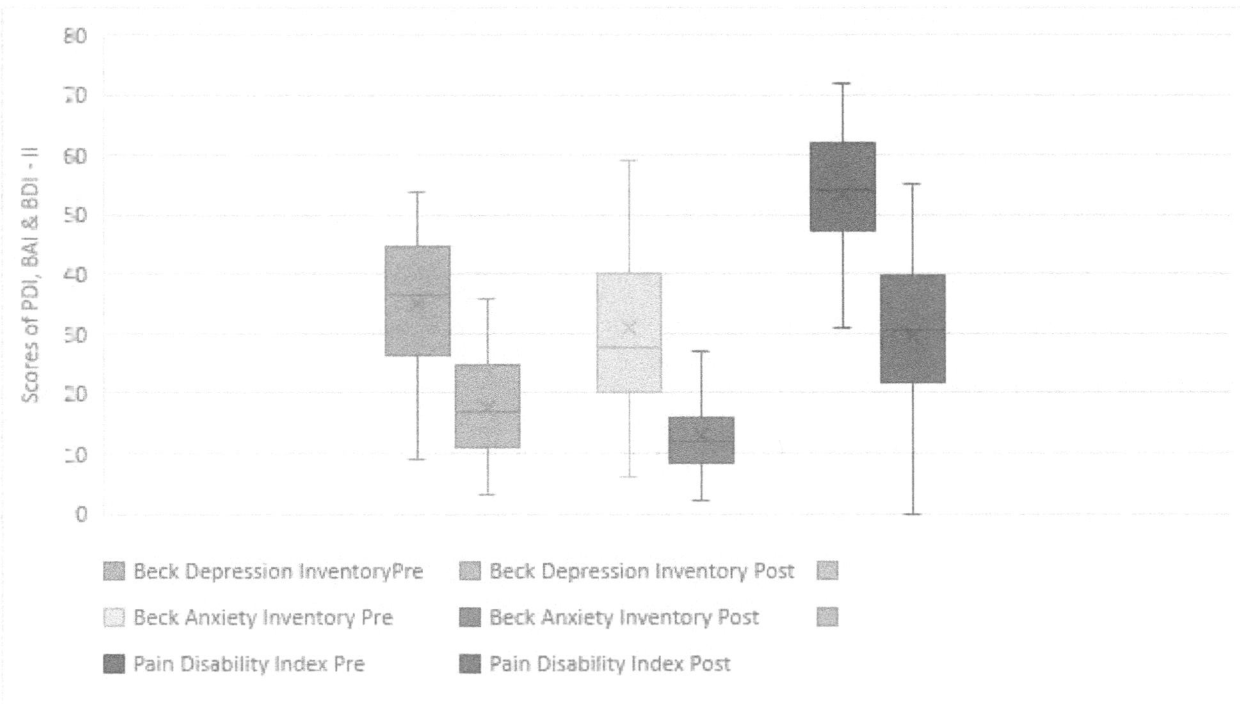

Figure 4.2.2.3. Pre-intervention and post-intervention scores of pain disability index, beck anxiety inventory and beck depression inventory-II of participants of aerobic exercise group.

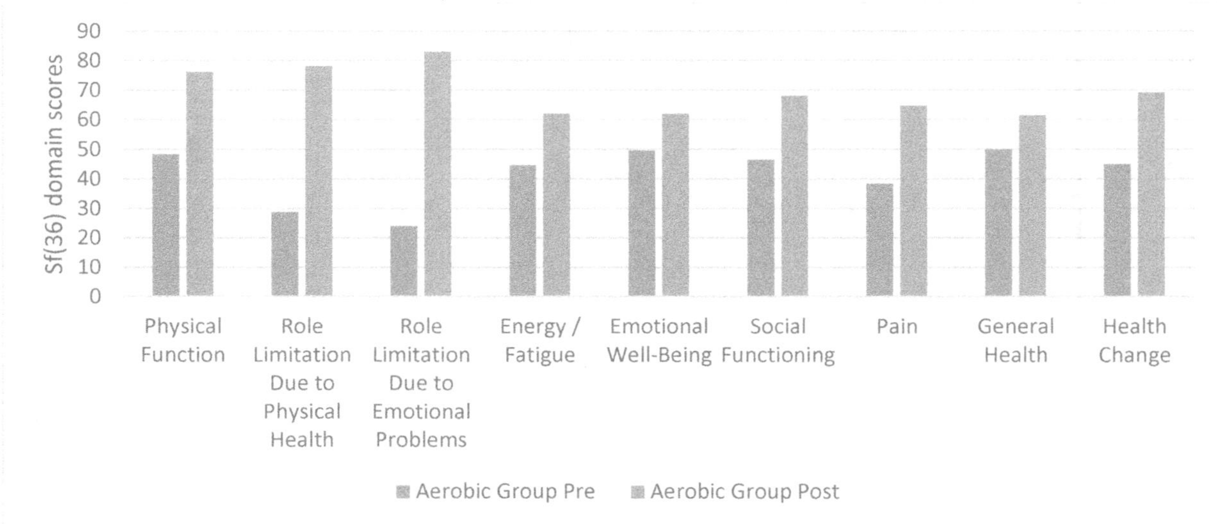

Figure 4.2.2.4. Pre-intervention and post-intervention scores of domains of Short form 36 (Sf36) for aerobic exercise group.

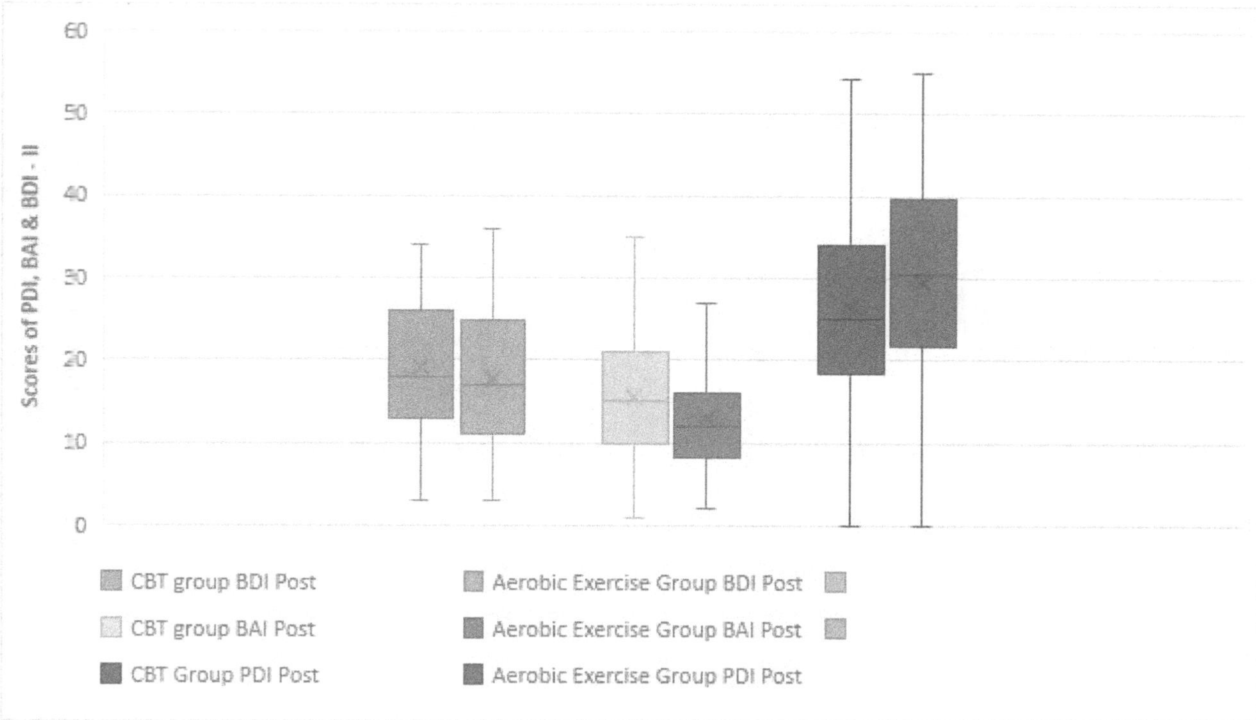

Figure 4.2.3.1. Between the group comparison of Post-intervention scores of pain disability index, beck anxiety inventory and beck depression inventory-II of CBT and aerobic exercise groups.

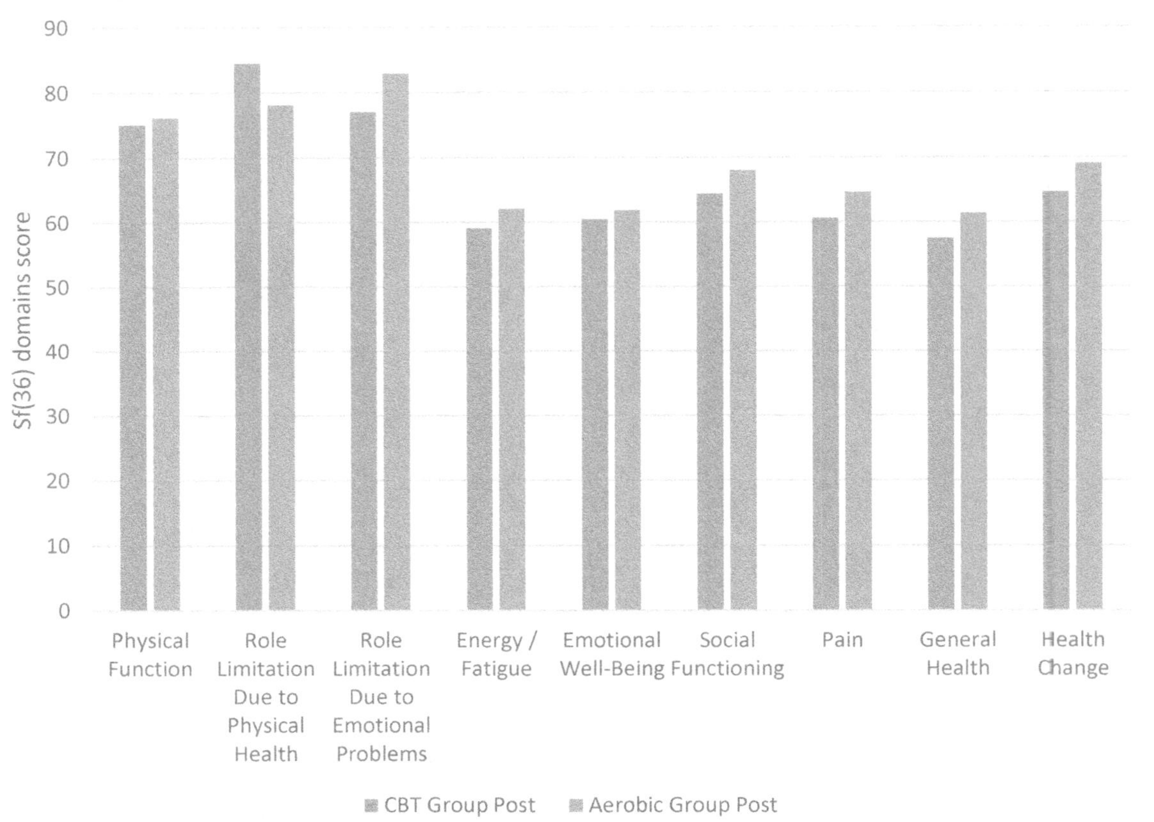

Figure 4.2.3.2. Between the group comparisons for post-intervention scores of nine domains of Sf (36) of CBT and aerobic exercise groups.

4.2.4. Effect size of CBT and aerobic exercise interventions

The effect size for CBT and aerobic exercises on participants having PMS has been determined by working out Cohen's d. according to the expression given in section 3.14. of the chapter 3 on material methods. The Cohen's d scores of CBT group for pain disability index, beck anxiety inventory and beck depression inventory –II is shown in the table 4.2.4.1. The Cohen's d scores for the parameters after CBT intervention are 1.48, 1.37 and 1.63, respectively suggesting that CBT intervention is more useful in reducing depression caused by PMS.

Table 4.2.4.1. Effect Size of CBT intervention on pain disability index, beck anxiety inventory and beck depression inventory-II.

Outcome measure	CBT (Mean ± SD) group Pre-intervention	CBT (Mean ± SD) group Post-intervention	95% Confidence interval	t-value	p-value	Cohen's d
Pain Disability Index (PDI)	51.39 ±20.02	26.92 ± 12.19	0.91, 2.05	8.09	<0.001	1.48
Beck Anxiety Inventory (BAI)	30.67 ±13.84	15.48 ±7.93	0.79, 1.91	7.38	<0.001	1.35
Beck Depression Inventory-II (BDI-II)	34.99 ±10.83	19.03 ± 8.58	1.05, 2.22	8.09	<0.001	1.63

The Cohen's d values for pain disability index, beck anxiety inventory and beck depression inventory –II for aerobic exercise group is shown in the table 4.2.4.2. The Cohen's d values for pain disability index, beck anxiety inventory and beck depression inventory –II was 1.86, 1.70 and 1.70, respectively which indicate that the reduction in scores for these outcome measures after aerobic exercise intervention is large and clinically meaningful with aerobic exercises showing maximum effect on reducing pain.

Table 4.2.4.2. Effect Size of aerobic exercise intervention scores on pain disability index, beck anxiety inventory and beck depression inventory-II.

Outcome measure	Aerobic exercise group (Mean ± SD) Pre-intervention	Aerobic exercise group (Mean ± SD) Post- -intervention	95% Confidence interval	t-value	p-value	Cohen's d
Pain Disability Index (PDI)	52.72 ± 12.48	28.95 ± 13.06	1.14, 2.58	13.44	<0.001	1.86
Beck Anxiety Inventory (BAI)	30.85 ± 13.83	12.88 ± 5.76	1.00, 2.39	12.82	<0.001	1.70
Beck Depression Inventory-II (BDI-II)	35.33 ± 12.08	17.61 ± 8.45	1.01, 2.39	12.15	<0.001	1.70

The Cohen's d values of CBT group for Sf (36) domains are given in table 4.2.4.3. The effect size for Sf (36) domains for cognitive behavior therapy intervention varied from 0.83 for general health to 2.68 for role limitation due to physical health followed by 1.82 for role limitation due to emotional problems. The ascending effect size of 1.11, 1.18, 1.33, 1.36, 1.58, 1.64 and 2.68 was seen for emotional wellbeing, energy/fatigue, physical function, social function, pain and role limitation due to physical health, respectively.

Table 4.2.3.3. Effect size of CBT intervention on nine domains of Sf (36)

Sf (36) Domains	CBT group (Mean ± SD) Pre-intervention	CBT group (Mean ± SD) Post-intervention	95% Confidence interval	t-value	p-value	Cohen's d
Physical Function	46.53 ± 22.29	75.14 ± 19.73	-1.92, -0.8	-11.38	<0.001	1.36
Role Limitation Due to Physical Health	28.13 ± 21.97	84.51 ± 20.10	-3.37, -1.98	-15.69	<0.001	2.68
Role Limitation Due to Emotional Problems	26.22 ± 24.04	77.06 ± 31.46	-2.42, -1.21	-10.44	<0.001	1.82
Energy/ Fatigue	45.69 ± 11.51	59.06 ± 11.18	-1.73, -0.63	-7.72	<0.001	1.18
Emotional Well-Being	47.28 ± 11.36	60.47 ± 12.44	-1.65, -0.56	-7.72	<0.001	1.11
Social Functioning	43.34 ± 12.22	64.43 ± 14.44	-2.16, -1.00	-9.39	<0.001	1.58
Pain	37.10 ± 10.11	60.59 ± 17.56	-2.22, -1.05	-11.52	<0.001	1.64

General Health	47.50 ± 13.10	57.50 ± 10.90	-1.36, -- 0.30	-6.77	<0.001	0.83
Health Change	39.86 ± 19.76	64.69 ± 17.61	-1.89, - 0.77	-7.58	<0.001	1.33

The effect size for aerobic exercise group for Sf (36) domains is given in table 4.2.4.4. The Cohen's d values over the nine domains of Sf (36) varies from 0.81 to 2.33 with the largest effect size of 2.33 for role limitation due to emotional problems followed by role limitation due to physical health with Cohen's d value of 1.72. Smallest effect size of 0.81 was seen for general health.

Table 4.2.4.4. Effect Size of aerobic exercise intervention on Sf (36) domains

Sf (36) Domains	Aerobic exercise group (Mean ± SD) Pre-intervention	Aerobic exercise group (Mean ± SD) Post-intervention	95% Confidence interval	t-value	p-value	Cohen's d
Physical Function	48.43± 21.87	76.19 ± 15.75	0.94, 1.97	-11.06	<0.001	1.46
Role Limitation Due to Physical Health	28.82 ±28.66	78.13 ± 28.43	1.16, 2.30	-10.28	<0.001	1.73
Role Limitation Due to	24.09 ±24.74	82.93 ± 26.27	-1.61, 2.99	-13.12	<0.001	2.33

Emotional Problems						
Energy / Fatigue	44.72 ±11.53	62.08 ± 10.86	1.01, 2.09	-9.57	<0.001	1.55
Emotional Well-Being	49.51 ±12.50	61.88 ± 12.08	0.56, 1.45	-8.26	<0.001	1.01
Social Functioning	46.46 ±12.30	68.02 ± 15.36	1.01, 2.09	-9.28	<0.001	1.55
Pain	38.39 ±18.68	64.65 ± 18.03	0.92, 1.94	-8.82	<0.001	1.43
General Health	50.00 ±16.61	61.39 ± 10.88	0.40, 1.23	-5.65	<0.001	0.81
Health Change	44.97 ±19.04	69.07 ± 20.82	0.73, 1.68	-9.04	<0.001	1.21

A comparison of the effect size for CBT and aerobic exercises on alleviation of PMS assessed using Cohen's d is given in table 4.2.4.5. The Cohen's d score for CBT and aerobic exercise was ----, respectively. Although, the overall improvement in all 12 parameters assessed is almost same after CBT and aerobic exercises, maximum improvement was seen in role limitation due to physical health after CBT intervention and the role limitation due to emotional problems.

Table 4.2.4.5. Comparison of effect size for CBT and aerobic exercise groups

Sr.No.	Outcome measure/Domains	Cohen'd CBT group	Cohen'd aerobic exercises group

1.	Pain Disability Index	1.48	1.86
2.	Beck Anxiety Inventory	1.35	1.70
3.	Beck Depression Inventory-II	1.63	1.70
Sf (36) Domains			
1.	Physical Function	1.36	1.46
2.	Role Limitation Due to Physical Health	2.68	1.73
3.	Role Limitation Due to Emotional Problems	1.82	2.33
4.	Energy/ Fatigue	1.18	1.55
5.	Emotional Well-Being	1.11	1.01
6.	Social Functioning	1.58	1.55
7.	Pain	1.64	1.43
8.	General Health	0.83	0.81
9.	Health Change	1.33	1.21

5. DISCUSSION

The objectives of current study were to evaluate the effectiveness of CBT and aerobic exercises in pain disability index, beck anxiety index, beck depression inventory-II and RAND Sf (36) and compare and evaluate the effect of cogitative behavioral therapy and aerobic exercises in mitigation of premenstrual syndrome symptoms in adolescent girls. This study employed an experimental design with sample size of 144 adolescent girls, divided equally to receive either CBT or participate in aerobic exercises over a period of 08 weeks. The findings of this study indicated notable variations in age of onset of mensuration in girls. The mean age of menarche was found to be 13.01±1.22 years. The majority of 31.25% girls experienced menarche at age of 14 years, followed by 36.81% of girls at age of 13 years, then 15.97% girls at 12 years and at the age of 10 years 3.47% and 8.33% had menarche at 15 years, only 0.69% girls had their menarche at the age of 17 years. These results align with study of Rai et al., (2020) who reported a similar distribution of menarche age, was 13.50 ±1.30, likewise, the study of Jami et al., (2024) reported the first age of menstruation in their study was 12 years. Whereas, Williams et al., (2019) observed an earlier onset of menarche and delayed menarche at the age of 15 years was uncommon. It is likely due to regional and genetic differences. The late menarche is due to high physical activity, chronic malnutrition, and psychological stress. The higher menarche in this study might be due to adequate nutrition and fewer physical stressors. Additionally, the findings were supported by Martinez et al., (2018) who suggested that socioeconomic and education status have positive impact on menarche age. The environmental exposures might accelerate puberty. These variations highlight the complex interplay of genetic, nutritional, socioeconomic, and environmental factors in determining menarche age (Biro et al., 2018). Author of present study found that early menarche might be due to the lifestyle, lack of physical activities, food habits and more screen time.

The BMI of the 54% population of the study was normal, while 31% were classified as underweight, 12% as overweight, and 3% as obese. The results are in contrast with the study of Williams et al., (2019) who observed a lower prevalence of underweight individuals, potentially due to regional, dietary variations and socioeconomic factors. The higher proportion of underweight participants may be attributed to nutritional deficiencies, lifestyle factors or genetic predispositions. Normal BMI participants in present study may be due to awareness of good health, socioeconomic status of the participants and education levels of the parents and participants.

The length of menstrual cycle of most of adolescent girls (70%) were of 30 days, which is in line with existing literature of Brown et al., (2016), who noted that the majority of adolescents tend to develop regular cycles within a few years of menarche. However, the proportion of girls reporting cycle lengths of 28 days were 12.22% and 25 days were 6.67% in this study. It was slightly higher than compared to some other studies, possibly due to regional and geographical variations, hormonal influences, or differences in sample characteristics as it aligns with the study of Martin et al., 2018, it was reported that 70.6% females had cycles in between 22 to 28 days of cycle.

Around 80.65% participants of this study, experienced moderate flow, which is consistent with research of Beevi et al., (2017). The heavy flow was reported in 18.28% which could be due to variations in factors like stress, nutritional status, or genetic predispositions that affect menstrual health. Additionally, a very small proportion of girls (1.08%) were having scarce flow. This might be due to a subset of adolescents experiencing irregularities due to hormonal imbalances or other underlying health conditions. These differences could be attributed to regional, environmental, or lifestyle factors influencing menstrual health. Similarly, Omidvar et al. (2019) reports that 69.9% girls had scarce to moderate blood flow whereas, 30.1% females had abundant blood loss during menstruation.

The majority of adolescent girls of present study were found to be well verse with menstrual hygiene. 75% of girls reported that they were changing their absorbent every 6 hours because of long hours of college or clinical postings or they are busy in the care of patients and only 26% reported changing every 4 hours. The girls use college washrooms to change their sanitary napkins and few go back to their hostel to change. The results are aligned with the findings of Brown et al., (2016) and Martinez et al., (2018) who suggested personal preference for frequently changing of absorbent to maintain hygiene. While only 1% of participants reported 10-hour interval between two changes. This may also reflect individual preferences or varying levels of access to sanitary products.

All participants (100%) exclusively used commercial pads, which is in contrast with findings of Ashan Biju, (2023). During a project of UN, he observed that only 36% of Indian women are using pad and the trend is on rise in teenagers for use of commercial pads. Study conducted by Borkar et al., (2022) found that around 73.79% of girls were using sanitary pads, while 26.21% of girls were using cloth. It is the most commonly used menstrual product among adolescent girls. Study conducted by van Eijk et al. (2016) reported, pooled estimation of pad users was 45% which was higher than that of a report of 2010, 1033 females were sampled across India and only 15% of them were using pads. The absence of alternative products such as tampons or menstrual cups in this study may be influenced by cultural preferences, lack of awareness, or availability of these products in the study region.

All participants (100%) of present study were following proper cleaning practices, which is encouraging and aligns with recommendations from ministry of drinking water and sanitation, India on menstrual hygiene management. This high adherence to hygiene practices may be due to increased awareness, medical education around menstrual health in the study population, similar trend has been reported by Biro et al., (2018) as well.

In the present study, the PSST-A scores revealed that the majority (96.32%) reported moderate symptoms of PMS, while only 3.68% experienced severe symptoms. These findings were advocated by Jadhav et al., (2023) who stated that 35.6% of females were suffering from PMS, distributed as 45% of them had mild symptoms, 36.6% had moderate and 22.4% had severe PMS. 40.9% of urban girls and 51.6% of rural girls were suffering from premenstrual syndrome reported by Rupvani, (2014). Whereas Balaha et al., (2010) reported that 70% of the respondents reported having dysmenorrhea. Out of these, 15.72% experienced severe pain, 34.2% experienced moderate pain, and 50% experienced mild pain. The relatively low proportion of girls in present study showed moderate symptoms which may suggest that the severity of PMS in this population tends to remain on the milder end, possibly due to factors like early age, hormonal fluctuations, or lifestyle.

The PMMS scores indicated that the majority (99%) experienced mild symptoms of PMS, whereas only 1% reporting moderate symptoms. No girl reported severe PMS symptoms, which align with findings from Brown et al., (2016), who also observed a predominance of mild symptoms in adolescent populations. The absence of severe symptoms could be attributed to factors such as the relatively young age of the participants, which may lead to less intense symptomatology or the possible underreporting of more severe cases. These findings highlight the predominance of mild PMS symptoms in adolescent girls. A study of Kamat et al., (2019) reported mild to moderate premenstrual symptoms and anger or irritability was most common symptom among all, followed by fatigue or lack of energy and by decreased interest in work activity.

The prevalence of premenstrual syndrome among college girls of this study comes out to be 81.11% where, a total of 157 students were examined. Among various studies conducted in other part of India reported different prevalence rate like Singh et al., (2008) reported that 60.5% of the medical students suffered from PMS, Lakshmi *et al.*, (2011) found the prevalence of PMS to be 67% whereas various studies reported prevalence of PMS as 55.8%, Ramya et al., (2014) reported the prevalence of PMS in urban girls as 40.9% and in rural girls as 51.6% and Sarkar *et al.*, (2015) reported prevalence of PMS as 61.5%, Upadhyay et al., (2023) reported the prevalence of PMS was 86%, a high prevalence rate of PMS was also reported by another study done by Nandakumar et al., **(**2023) which showed that 76.35% students had PMS. The prevalence of PMS was considerably greater among people who were between the ages of 18 - 21 and were hostellers. Nearly similar results were demonstrated by Kamat et al., (2019) who revealed that 95% prevalence rate in adolescent girls. Tarannum F et al., (2022) reported prevalence of PMS as 37.**7%.** In contrast to some of the studies conducted outside India by Abeje et al., (2016) and Shiferaw et al.,

(2014) found the prevalence of PMS in University students to be 81.3% and 72.8% respectively in Ethiopia. Two studies conducted in Sistan of Iran in Baluchistan University and Zahedan University, Naeimi, (2015) and Bakhshani et al., (2009) respectively reported a lower prevalence of 36.3% and 16%, respectively, as compared to the present study. This variation in prevalence rate can be because of difference in age group of the study, study design, study methodology, sample size, difference in cultural background and geographical variation.

In the present study, the comparison of pre and post-intervention scores of Pain Disability Index (PDI) for Cognitive Behavioral Therapy (CBT) showed a significant improvement. The mean PDI score decreased from 51.39 ± 20.0 (pre-intervention) to 27.92 ±12.19 (post-intervention), with a t-value of 8.09 and a p-value of <0.001. This finding indicates a highly significant reduction in pain-related disability following the intervention. The substantial decrease in the PDI score suggests that CBT can be an effective treatment modality which aligns with the study of Kennedey et al., (2014) for improving the quality of life and reducing the impact of pain-related disability in individuals with PMS. Similar study conducted by Maddineshat et al., (2016) showed that CBT was effective on the psychological symptoms of PMS such as anxiety, depression, negative thoughts and physical changes. The results of this study were in line with meta- analysis conducted by Relly et al., (2015) and the similar result was also seen in a randomized controlled trial conducted by Kues et al., (2014), where a standardized, manualized intervention was given to subjects and showed positive results. In reducing pain CBT is effective; as pain is consider having psycho-social component. While reflecting on the underlying negative thoughts, CBT helps in identifying negative thoughts and helps in challenging belief about pain which can reduce pain perception. CBT helps in behavioral modification increased acceptance of pain increased and hence threshold of pain also increases. Nazari et al., (2012) advocated that by avoiding worrying thoughts, training creative thoughts, developing self- believing, accepting both abilities and shortcomings, reducing expectancies from oneself and the others, improving purposefulness by planning and organizing activities could relieve anxiety from PMS.

The SF (36) scores for Group CBT demonstrated significant improvements across all components after the intervention. For physical functioning, the mean value increased from 46.53 ± 22.29 (pre-intervention) to 75.14 ± 19.73 (post-intervention), with a t-value of -11.38 and p-value of <0.001. Similar improvements were observed in role limitations due to physical health, emotional problems, energy/fatigue, emotional well-being, social functioning, pain, general health, and health change, all showing very significant results. These results highlight substantial enhancements in both physical and mental health-related quality of life for Group CBT participants. The findings indicate that CBT is an effective intervention to improve the health-related quality of life of women with PMS. The findings of this study replicate the results of previous studies Davoodvandi et al., (2011); Aruna (2013); Taghizadeh et al., (2013); Maryam M et al., (2016); Ussher et al., (2002); Hunter et al., (2002) and Blake et al., (1998), which suggested that the CBT could be effective for women with PMS. The previous researches demonstrated the positive impact of CBT on improving health-related quality of life in individuals with chronic conditions. Present study suggest that CBT is effective in enhancing both physical and psychological well-being.

Group CBT experienced a significant reduction in anxiety levels following the intervention. The mean anxiety score decreased from 30.67 ± 13.84 (pre-intervention) to 15.48 ± 7.93 (post-intervention), with a t-value of 7.38 and p-value of <0.001. This substantial decrease reflects a marked improvement in anxiety symptoms, indicating the effectiveness of CBT in alleviating anxiety. These results supported by Faeze P and Mahbobeh F, (2016), highlighted that mindfulness - based CBT as an effective intervention for reducing anxiety levels in individuals with chronic conditions. The significant reduction in anxiety in this study also suggests that CBT can play a crucial role in improving mental health and well-being of the female who is suffering from PMS. Blake et al., (1998) observed that CBT can reduce psychological but not somatic symptoms in women with PMS. Contrary report of Morse et al., (1991) shows that symptoms of anxiety, depression and PMS can be improved by progesterone and relaxation but not with CBT.

The Beck Depression Inventory-II scores for Group CBT showed significant improvements, with the mean score dropping from 35.1 (pre-intervention) to 17.6 (post-intervention). The t-value of 14.9 and a p-value of <0.001 indicate a considerable reduction in depression symptoms following the intervention. These findings suggest that CBT is highly effective in alleviating depressive symptoms, in line with previous research supporting CBT's role in improving mental health outcomes for individuals with PMS Kues J et al., (2014). The marked reduction in depression symptoms highlights CBT as a powerful tool for enhancing emotional well-being. Lustyk M et al., (2009) supported the use of CBT to treat PMS, according to the author the core focus of CBT is to change the perception of an individual about that particular event which can affect the individual in certain way or in which way individual is experiencing that event which include physical sensations. Hence CBT is effective in treating mood and anxiety disorders and shown helpful in reliving physical symptoms. The relieving of the symptoms of anxiety and depression may help women with PMS to manage their emotions better and have a core therapeutic impact.

The comparison of pre- and post-intervention pain disability index (PDI) scores for group aerobic exercise revealed a significant improvement. The mean score decreased from 51.39 ± 20.02 (pre-intervention) to 26.92 ± 12.19 (post-intervention), with a t-value of 8.09 and a p-value of <0.001. This substantial reduction in pain-related disability highlights the effectiveness of aerobic exercise in alleviating pain and improving functioning. These results are advocated by Pilozzi A et al., (2021) who suggested that aerobic exercise of moderate intensity has demonstrated an increase in circulating beta endorphins. That could be the possible mechanism that is responsible for reduction in pain related symptoms among females suffering from PMS. These findings are in accordance with previous research of Samadi et al., (2013) that has demonstrated the positive impact of aerobic interventions in reducing physical symptoms. The significant improvement in PDI scores suggests that aerobic exercise can be a valuable intervention for managing chronic pain caused by PMS. Highly physically active females had lower physical and psychological symptoms of PMS as it was suggested by Kawabe et al., (2022).

The SF (36) scores for group comprising aerobic exercises showed significant improvements across all components following the intervention. For physical functioning, the mean value increased from 48.43 ± 21.87 (pre-intervention) to 76.19 ± 15.75 (post-intervention), with a t-value of -11.06 and a p-value of <0.001. Role limitations due to physical health and emotional problems also demonstrated notable improvements, with pre- to post-intervention means increasing from 28.82 ± 28.66 to 78.13 ± 28.43 and from 24.09 ± 24.74 to 82.93± 26.27 respectively, with p-value <0.001. Additional components, including energy/fatigue, emotional well-being, social functioning, pain, general health and health change also showed highly significant improvements (p-values <0.001). These findings highlight considerable enhancements in both physical and mental health-related quality of life for group Aerobic exercise. The consistent improvements across various SF-36 domains suggest that aerobic exercise intervention can be highly effective in promoting comprehensive health benefits for individuals with chronic conditions. The findings of this study was similar to the study of Eda A and Erdem E, (2019) reported that Four-week group aerobic training program can effectively reduce premenstrual symptoms and improve quality of life in females with primary dysmenorrhea. Author suggested that aerobic activity increase the perception of wellness, reduce stress and lower the symptoms of PMS, as brain releases chemicals such as endorphins which brings happiness and relief to the individual. The results of systematic review of Ayyub S et al., (2024) indicated that the severity of premenstrual disorder symptoms can be effectively reduced by the physical activities which are also supported by the recommendation of clinical guidelines to consider physical activity as a primary prevention and treatment for PMS, alongside with other management strategies. On similar grounds, the study of Pazoki H et al., (2015) concluded that fennel extracts and aerobic exercise are effective in reducing the intensity of premenstrual syndrome in young women. The result also shows that fennel extract has more significant effects. After the intervention of 8 weeks' aerobic exercise along with fennel extracts showed reduced intensity of PMS. Hence, the fennel extract and aerobic exercise could be used together to treat women with PMS.

Aerobic exercise group exhibited a significant reduction in anxiety levels following the intervention. The mean score decreased from 30.85 ± 13.83 (pre-intervention) to 12.88 ± 5.76 (post-intervention), with a t-value of 12.82 and p-value of <0.001. This substantial decrease reflects a strong reduction in anxiety symptoms, demonstrating the effectiveness of aerobic exercise in alleviating anxiety. These results align with previous study of Guszkowska M et al., (2004) that have highlighted the positive impact of aerobic interventions on reducing anxiety levels and improving mental well-being. The results confirm that there is reduction in anxiety and depression after single session of 15-30 minutes of aerobic exercises involving large muscle group. Rani K et al., (2011) advocated the impact of Yoga Nidra on psychological general well-being in patients with menstrual irregularities. The study reveals that anxiety decreased significantly with Yoga Nidra. This may be due to endorphin and monoamine hypothesis. Exercise increases the body temperature, blood circulation and impact on hypothalamic-pituitary-adrenal axis and physiological reactivity to stress. These possible psychological mechanisms also improve self-efficacy, reduce distraction and cognitive dissonance. The significant improvement in anxiety suggests that aerobic exercise can be a valuable tool for promoting psychological health in individuals with chronic conditions due to PMS.

The Beck Depression Inventory-II scores for group comprising aerobic exercise also showed significant improvements. The mean score dropped from 35.33 ± 12.08 (pre-intervention) to 17.61 ± 8.45 (post-intervention), with t-value of 12.15 and a p-value of <0.001. This reflects a notable reduction in depression symptoms following the intervention, highlighting the effectiveness of aerobic exercise in alleviating depressive symptoms. Similar findings were reported by Udatha T et al., (2020) who demonstrated that aerobic exercise had helped in reducing the depressive symptoms, anxiety and improved the self-esteem and quality of life of patients with depression. Author suggested that it is a cost-effective and feasible method to be used along with medication for patients to manage depressive symptoms. Rani K et al., (2011) emphasizes that Yoga helps in decreasing the cortisol levels which leads to a counter-regulatory effect hence, reliving depression. There is positive impact of aerobic exercise on reducing depression levels and enhancing overall mental health. Dunn A et al., (2001) and Pearce E et al., (2020) reported that physical exercise has the ability to elevate endorphin levels, regulate the synthesis of progesterone and estrogen, and stimulate the production of naturally occurring anti-inflammatory substances. The exercise offers various other advantages, including enhanced overall fitness, self-confidence for social interaction and reduces the depressed feelings. These combined benefits may contribute to moderating the range of symptoms experienced in PMS. According to research, engaging in physical activity lowers renin activity and raises estrogen and progesterone levels, which in turn lowers serum aldosterone levels and the reabsorption of water and salt. This reduces edema and alleviates physical symptoms. Joyner and Charkoudian (2004), revealed that aerobic exercise helped women's estrogen and progesterone levels balance, which in turn lessened symptoms. The physical activity reduces the symptoms of PMS as it has been advocated by the result of study done by Kawabe R et al., (2022) and Ravichandran et al., (2022).

The post-intervention comparison of PDI scores reveals no significant difference between group CBT and group aerobic exercise. The mean scores are very similar, with group CBT at 26.92 ± 12.19 and group aerobic exercise at 28.95 ± 13.06. The t-value of -0.62 and p-value of 0.53 indicates that this difference is not statistically significant. This suggests that both interventions were equally effective in reducing pain-related disability in PMS. These finding implies that both CBT and aerobic exercises can provide comparable improvements in pain management, offering flexible options for individuals seeking to reduce pain-related disability. The results of this study demonstrate that both CBT and aerobic exercise significantly improved various aspects of health-related quality of life, with a few notable differences between the two interventions. Group CBT and group aerobic exercise shows similar improvements in most domains of the Sf (36), such as physical function, role limitations due to physical health, role limitations due to emotional problems, emotional well-being, social functioning, pain, and health change. This suggests that both interventions can effectively enhance overall quality of life, regardless of the specific intervention type. Bhuvaneshwari K et al., (2019) reported poor quality of life among students with PMS. Lifestyle factors also contribute to severity of PMS symptoms such as caffeine intake, sleep and physical activity. The result of present study can be compared to the results of Sharma et al., (2008) that reported, the most prevalent issue was dysmenorrhea (67.2%), while 63.1% of people experienced one or more PMS symptoms. Interestingly, the energy/fatigue domain, while showing a trend toward better scores in group aerobic exercise, did not reach statistical significance. This could be attributed to the complicated nature of fatigue, where factors like psychological stress, sleep patterns, and comorbid conditions may also play significant roles in influencing fatigue levels. Although aerobic exercise is typically associated with improvements in energy, the lack of a statistically significant finding may suggest that the intervention period was not long enough, or the sample size may not have been sufficient to detect this subtle effect. In terms of pain and physical function, both interventions produced significant reductions in pain-related disability, with no substantial difference between the two groups. This suggests that both CBT and aerobic exercise are effective in managing pain, which is consistent with findings from other studies that highlight the benefits of psychological and physical interventions in chronic pain management. Brown et al., (2017) summarize that both CBT and aerobic exercise may address pain from different angles, CBT targets cognitive and emotional responses to pain, and aerobic exercise provides a physical approach in improving musculoskeletal health and reducing pain sensitivity. The lack of significant differences in role limitations due to emotional problems and social functioning supports the idea that both CBT and aerobic exercise are helpful in enhancing mental well-being and daily functioning. To address emotional issues such as anxiety and depression CBT can be recommended, aerobic exercise has shown to have a positive impact on mood and social interactions.

The significant reduction in anxiety and depression scores in both groups highlights the potential of both interventions in improving mental health. Previous studies have consistently found that both aerobic exercise and CBT are effective in reducing symptoms of anxiety and depression. Mahmoudi F et al., (2018) and Dehnavi et al., (2018) recommends 8 weeks of aerobic exercises combined with relaxation techniques helps in reducing symptoms of PMS, anxiety, depression and negative thoughts. These findings emphasize on the importance of addressing both physical and psychological aspects of well-being in individuals experiencing PMS.

The post-intervention beck anxiety inventory scores for both group CBT and group aerobic exercise indicate no significant difference between the two groups, with group CBT showing a mean score of 15.48 ±7.3 and group aerobic exercise showing a slightly lower mean score of 12.88 ± 5.76 with t-value 1.45 and p-value 0.15, this confirms that the difference in anxiety levels between the groups is not statistically significant. Both interventions were similarly effective in reducing anxiety levels. Cognitive patterns and behaviors that leads to anxiety can be reduced by CBT, while aerobic exercise is helpful in enhancing mood and reducing stress through physiological mechanisms like the release of endorphins. Despite the different approaches, both interventions appear to have a comparable impact on anxiety reduction, which aligns with other research done by Brown et al., (2017) which suggest that both psychological and physical interventions can effectively lower anxiety symptoms. Aerobic exercises are effective in improving premenstrual symptoms it was recommended by Ravichandran H and Janakiraman B, (2022). This review is evident for improving physical and physiological symptoms among women with pre-menstrual syndrome. Similar findings were seen in Saglam et al., (2020) where the authors stated that exercise helps with psychological symptoms like anxiety, depression and physical symptoms like pain, constipation and breast sensitivity. The effectiveness of both interventions in reducing the symptoms that targeting the underlying psychological mechanisms of anxiety, stress and depression can be cope up through cognitive restructuring and coping strategies (CBT) or through physical activities benefits on stress reduction and mood enhancement.

Furthermore, the absence of any statistically significant difference may indicate that both interventions are valuable options for individuals with anxiety and the choice of coping intervention between CBT and Aerobic exercise could depend on personal preferences, accessibility and other individual factors. Summarizing these results, it supports that both CBT and aerobic exercise are viable approaches for reducing anxiety, stress and depression providing individuals with multiple options for managing their symptoms.

The post-intervention beck depression inventory-II scores for group CBT (mean = 19.03 ± 8.58) and group aerobic exercise (mean = 17.61± 8.45) does not show any significant difference ($t = 0.65$, $p = 0.52$), indicating both interventions had similar effects in reducing depressive symptoms. Cuijpers et al., (2016) suggested that identifying and changing negative thought patterns can be done by CBT and aerobic exercise has its mood-enhancing physiological effects, as both are known for their efficacy in treating depression. Both the interventions are different in their approaches, the comparable outcomes suggest that both offer valuable avenues for managing depression, potentially giving individuals the flexibility to choose based on preference or specific needs. As there is no statistical significant difference could also reflect the fact that both interventions target key aspects of depression, either through cognitive and emotional regulation or through biological effects of physical activity, that includes improved serotonin and endorphin levels. These findings align with previous studies by Mann et al., (2015) and Wipfli et al., (2011) which showed that physical exercise can have therapeutic effects comparable to psychological interventions in treating depressive symptoms.

The findings of present study reveals that both Cognitive Behavioral Therapy (CBT) and aerobic exercise are highly effective in reducing anxiety, depression, pain-related disability, and improving health-related quality of life (HRQoL), it is consistent with a large body of research. Joshi A et al., (2017) and Vagehala N et al., (2019) showed positive impact of regular aerobic exercise on reducing perceived stress in luteal phase. It also found that aerobic exercise group showed a statistical significant decrease in perceived stress. Beta- endorphin escalates with aerobic exercises and increase in beta-endorphins enhance mood as well gives analgesics relief from pain. Exercises eliminate negative thoughts and enhance positive thoughts that also help in coping up with stress. It was further discussed that aerobic exercises intensely raise serum progesterone levels that gives positive result in mood alteration and reducing stress via neurotransmitter systems modulated by sex steroids. The large effect sizes observed for both the beck anxiety inventory and beck depression inventory (Cohen's d values 1.35 and 1.63 respectively) demonstrate that both interventions were highly effective in addressing anxiety and depression, aligning with findings from similar studies Hofmann et al., (2012) reported that CBT has a large effect size of 0.88 for treating anxiety and depression, supporting the effectiveness of CBT in addressing these conditions. Similarly, a study by Cuijpers et al., (2016) found that CBT is highly effective in treating depression with effect sizes around 0.88.

The effect sizes of group aerobic exercise oh this study for anxiety and depression is similar i.e. 1.70. This larger effect size highlights the strong impact of these intervention. These findings are similar with the study of Rebar et al., (2015), who reported moderate to large effect sizes (Cohen's d = 0.88 for anxiety and 0.77 for depression). Hence, the Cohen's d values observed in this study suggested that aerobic exercise can be a powerful intervention in managing mood disorders and at par with CBT.

In terms of pain-related disability, both group CBT and group aerobic exercise showed significant reductions, with Cohen's d values of 1.48 for group CBT and 1.86 for group aerobic exercise. These findings align closely with previous studies on CBT and aerobic exercise for pain management. Williams et al., (2012) found that CBT can lead to moderate to large reductions in pain-related disability. Additionally, aerobic exercise has been widely studied for its effectiveness in chronic pain management, Ang et al., (2015) demonstrated similar trend that aerobic exercise significantly reduced pain and improved physical functioning, with Cohen's d values ranging from 1.00 to 1.50.

The improvements in Sf (36), a health related quality of questionnaire (HRQoL) observed in both groups, particularly in physical functioning, role limitations due to physical health, and social functioning, are in accordance with the broader literature. Studies have shown that CBT not only improves mood but also enhances overall quality of life. For instance, a meta-analysis by Cuijpers et al., (2014) found that CBT significantly improves HRQoL, with moderate effect sizes (Cohen's d - 0.50 to 0.70). The study also found improvements in HRQoL, with group CBT exhibiting slightly larger effect sizes (Cohen's d -1.20) compared to group aerobic exercise (Cohen's d -1.00). This suggests that CBT may have a slightly greater impact on physical functioning, which is consistent with studies that have highlighted the importance of addressing physical health in chronic pain management. On the other hand, aerobic exercise is well-established as an effective means of improving physical health and quality of life. Research carried out by Schuch et al., (2016) has shown that exercise significantly improves physical functioning and quality of life in individuals with chronic conditions, which is consistent with the improvements observed in both groups of the study.

The reduction in role limitations due to physical health and role limitations due to emotional health in both groups supports the findings from previous research on both interventions. In present study, group CBT showed slightly higher Cohen's d values (d = 2.68) for role limitations due to physical health as compared to group aerobic exercise (d = 1.73). Both interventions were helpful in alleviating role limitations due to physical health, CBT intervention shows a marginal advantage in improving role limitations due to physical health. While, group aerobic exercise had slightly higher values for role limitations due to emotional health (d= 2.33) as compared to group CBT (d= 1.82) moderately greater impact on role limitations due to emotional health. This is in line with findings from other studies that have indicated that both psychological and physical interventions can help alleviate role limitations due to emotional health but may have varying effects depending on the nature of the limitation. Social function was improved in both the groups, indicating that both CBT and aerobic exercise had a comparable positive impact on participant`s social interactions and daily functioning. Previous study results also coincide with these results that have shown that both psychological and physical interventions can improve social functioning. CBT leads to significant improvements in social functioning, while aerobic exercise has been shown to improve overall well-being and social interactions and these findings were suggested by Hofmann et al., (2012) and Schuch et al., (2016). Both groups exhibit improvements in the domains of pain and general health, with Cohen's d values 1.43 and 0.81 respectively, similar study by Wipfli et al., (2011) found that exercise significantly improves health perceptions and reduces fatigue, coinciding with the results of current study. The group aerobic exercise showed more improvements in fatigue Cohen's d =1.55 and health change Cohen's d = 1.21.

This recommends that clinically aerobic exercise may have had a higher impact on betterment of perceived health status, which could be due to the physiological effects of exercise on mood, fatigue, and overall well-being. Thus, present study come to the conclusion that clinical impact of aerobic exercises are beneficial in reducing perceived stress, enhance mood by improving pain and physical functions. Instead, CBT act on psychological aspect of females hence, shows better clinical impact on improving quality of life. When CBT group and aerobic exercise groups were set side by side clinically for comparing the domains of quality of life, CBT shows improved domains of role limitation due to physical health, pain, emotional wellbeing and social functioning. On another hand showed significant clinical improvement in domains of physical function, role limitation due to emotional problems and energy /fatigue. General health and health change domains of Sf (36) showed marginal difference in both the groups.

SUMMARY

This study exhibited 81.11% prevalence of premenstrual syndrome (PMS) among medical students of the University which emphasize the need to cater effective interventions to this vulnerable population.

The result findings emphasize the effectiveness of CBT for 08 weeks on pain disability index and quality of life of adolescents. A significant reduction has been seen in the PID which indicates that CBT has been a potential modality and can be used to treat pain related disabilities occurred in PMS. CBT showed positive impact on health related quality of life. CBT well-acted on the various domains of Sf (36). Physical functioning, role limitation due to physical health, role limitation due to emotional problems, energy/fatigue, emotional wellbeing, social functioning, pain, general health and health changed has been improved after the intervention. It can be concluded that CBT can be used as effective treatment modality to treat symptoms of PMS and improve social involvement of the affected women.

Similarly, same pattern has been observed in the participants of group aerobic exercises. There has been substantial reduction observed in pain disability index after completion of 08 weeks' aerobic exercise sessions. Aerobic exercises reduce physical symptoms and hence can reduce disabilities caused by chronic pain. There has been significant improvement in quality of life of the participants of aerobic exercise group. The entire domains of Sf (36) like physical functioning showed significant improvement. Other domains of Sf (36) like, role limitation due to physical health, role limitation due to emotional problems, energy/fatigue, emotional wellbeing, social functioning, pain, general health and health changed also demonstrated good improvement. Hence, aerobic exercises can be administrated to relive symptoms of PMS.

When both interventions, CBT and aerobic exercises were compared to evaluate the clinical effectiveness of either the interventions, it was found that both the interventions were equally effective in treating symptoms of PMS. Aerobic exercise has shown more clinical effectiveness in treating disabilities caused by pain and improve physical functioning. Whereas CBT has shown slightly more clinical effect to improve anxiety, depression and health –related quality of life of the women suffering from PMS. CBT remarkably reduces psychological symptoms and aerobic exercises has good impact on physical symptoms of the PMS. This study conclude that both the interventions have modest difference, suggesting that both interventions are effective in their respective domains.

The findings from present study contribute to the growing body of evidence supporting the use of both psychological and physical interventions for managing pain-related disability, quality of life, anxiety and depression. Hence, either of the intervention can be used to improve PMS symptoms. The choice of intervention will be depending on the type and extent of symptoms of PMS. If psychological symptoms are more prevalent then CBT can be used and if physical symptoms are more prominent then aerobic exercises can be administrated. The aerobic exercise intervention is low cost intervention as compared to CBT, hence as per the requirement of the patient and choice of the patient either of the protocols can be used.

Conclusion

This study highlights that both cognitive behavioral therapy (CBT) and aerobic exercise can play a meaningful role in helping young women manage the challenges of Premenstrual Syndrome (PMS). CBT proved especially helpful in easing psychological symptoms like anxiety and depression, while also improving overall well-being and quality of life. On the other hand, aerobic exercise was particularly effective in reducing physical discomfort and pain-related disabilities. Both interventions showed improvements across various aspects of health and daily functioning. While each approach has its strengths, the best choice depends on the individual's specific symptoms. If emotional and psychological issues are more prominent, CBT might be the better option. If physical symptoms are more troubling, aerobic exercise may offer greater relief.

Importantly, aerobic exercise also stands out as a more affordable and accessible intervention, which could be especially beneficial in settings with limited resources. In conclusion, both CBT and aerobic exercise are valuable tools in supporting young women with PMS, and either can be chosen based on personal needs, preferences, and available resources.

References

Abootalebi, M., Khosravi, S., & Sharifi, Z. (2017). Cognitive behavioral nursing strategies for adolescent emotional regulation: Effects on premenstrual syndrome. Nursing Practice Today, 5(2), 104–112.

Abu Alwafa, R., Badrasawi, M., & Haj Hamad, R. (2021). Prevalence of premenstrual syndrome and its association with psychosocial and lifestyle variables: a cross-sectional study from Palestine. BMC women's health, 21(1), 233. https://doi.org/10.1186/s12905-021-01374-6

Acharya, A. (2006). The impact of premenstrual syndrome on adolescent quality of life. Journal of Adolescent Health, 32(4), 223–229.doi:10.1080/00016340600556049

Allais G, Castagnoli Gabellari I, Burzio C, Rolando S, De Lorenzo C, Mana O, Benedetto C. Premenstrual syndrome and migraine. Neurol Sci. 2012 May;33 Suppl 1:S111-5. doi: 10.1007/s10072-012-1054-5.

Anandha Lakshmi S, Priy M, Saraswathi I, Saravanan A, Ramamchandran C. Prevalence of premenstrual syndrome and dysmenorrhoea among female medical students and its association with college absenteeism. Int J Biol Med Res 2011;2:1011-6.

Antai, A.B.(2004). Premenstrual syndrome: Prevalence in students of the University of Calabar, Nigeria. Nigerian Journal of Medicine, 13(2), 146-150.

Arafa, M. (2018). Prevalence and impact of premenstrual syndrome among adolescents: A systematic review. Global Health Studies, 29(1), 75–84.

Ashritha, M., Rani, S. L., & Brundha, M. P. (2020). Effects of premenstrual syndrome in day-to-day life of women. Drug Invention Today, 14(7).

Bajalan, Z., Alimoradi, Z., & Moafi, F. (2019). Nutrition as a potential factor of primary dysmenorrhea: a systematic review of observational studies. Gynecologic and obstetric investigation, 84(3), 209-224.

Baker, F. C., Sassoon, S. A., Kahan, T., Palaniappan, L., Nicholas, C. L., Trinder, J., & Colrain, I. M. (2012). Perceived poor sleep quality in the absence of polysomnographic sleep disturbance in women with severe premenstrual syndrome. Journal of sleep research, 21(5), 535-545.

Bakhshani, N. M.,(2009). Association between dietary habits and premenstrual symptoms in adolescents. Journal of Nutritional Science and Dietetics, 7(1), 34–41.

Beck, A.T., Ward, C.H., Mendelson, M., Mock, J., & Erbaugh, J. (1961). An inventory for measuring depression. Arch Gen Psychiatry 4, 53–63.

Beckerman, H., Terwee, C. B., van der Windt, D. A., & Bouter, L. M. (2007). The measurement of quality of life in patients with chronic musculoskeletal disorders. Clinical Rehabilitation, 21(9), 866-877. https://doi.org/10.1177/0269215507079787

Bhalerao V, Gotarkar S, Muneshwar K and Vaishnav DL. A Cross-sectional study to assess the mental health problems during menstrual cycle among adolescent girls in the rural area of Wardha District [version 1; peer review: awaiting peer review]. F1000Research 2024, **13**:436 https://doi.org/10.12688/f1000research.146475.1

Biggs, W. S., & Demuth, R. H. (2011). Premenstrual syndrome and premenstrual dysphoric disorder. American family physician, 84(8), 918-924.

Blumenthal, J. A., Babyak, M. A., Doraiswamy, P. M., Watkins, L., Hoffman, B. M., Barbour, K. A.,& Sherwood, A. (2007). Exercise and pharmacotherapy in the treatment of major depressive disorder. Psychosomatic medicine, 69(7), 587-596.

Brahmbhatt, S., Sattigeri, B., Shah, H., Kumar, A., & Parikh, D. (2013). A prospective survey study on premenstrual syndrome in young and middle aged women with an emphasis on its management. International Journal of Research in Medical Sciences, 1(2), 69. https://doi.org/10.5455/2320-6012.ijrms20130506

Brown, C. H., Latham, C. E., & Hartsell, T. L. (2016). Interventions for chronic pain: Vitality and emotional health improvements. Pain Medicine, 17(9), 1592-1600. https://doi.org/10.1093/pm/pnw007

Brown, L. M., Green, S. T., & White, K. A. (2017). Heterogeneity in pain-related disability: Variability across chronic pain patients. Pain Management Nursing, 18(2), 83-89. https://doi.org/10.1016/j.pmn.2016.06.008

Cherkin, D. C., Sherman, K. J., Balderson, B. H., Cook, A. J., Anderson, M. L., Hawkes, R. J., Hansen, K. E., & Turner, J. A. (2016). Mindfulness-based stress reduction for chronic pain management: A systematic review and meta-analysis. Journal of Pain Research, 9, 257–270. https://doi.org/10.2147/JPR.S105748

Cicek, G. (2018). The effect of regular aerobic exercises on premenstrual syndrome in sedentary women. Baltic Journal of Health and Physical Activity, 10(2), 34–42. https://doi.org/10.29359/bjhpa.10.2.04

Ç til, S., and Kaya, H. (2019). Aerobic exercises and their role in managing premenstrual syndrome. Journal of Women's Health Care, 14(6), 678–685.

Connolly, M., O'Malley, E., & Peiris, A. T. (2001). Premenstrual syndrome: An update on definitions, diagnosis and management. Advances in Psychiatric Treatment, 7(6), 469–477. https://doi.org/10.1192/apt.7.6.469

Crofford, L. J., (2005). Fatigue and chronic pain: Management and implications for treatment. Clinical Journal of Pain, 21(4), 253-262. https://doi.org/10.1097/01.ajp.0000164457.41034.04

Cuijpers, P. (2008). Psychological treatment of depression in chronic pain patients: A meta-analysis. Journal of Consulting and Clinical Psychology, 76(6), 1023-1033. https://doi.org/10.1037/a0012708

Cuijpers, P., (2013). Psychological treatment of depression in patients with chronic pain: A meta-analysis of randomized controlled trials. The Journal of Pain, 14(3), 263-273. https://doi.org/10.1016/j.jpain.2012.11.004

Daley, A. (2009). Exercise and Premenstrual Symptomatology: A Comprehensive review. Journal of Women S Health, 18(6), 895–899. https://doi.org/10.1089/jwh.2008.1098

Dambhare, D. G., Wagh, S. V., & Dudhe, J. Y. (2012). Age at menarche and menstrual cycle pattern among school adolescent girls in Central India. Global journal of health science, 4(1), 105–111. https://doi.org/10.5539/gjhs.v4n1p105

Dehnavi, Z. M., Jafarnejad, F., & Goghary, S. S. (2018). The effect of 8 weeks aerobic exercise on severity of physical symptoms of premenstrual syndrome: a clinical trial study. BMC Women S Health, 18(1). https://doi.org/10.1186/s12905-018-0565-5

Derman, R., (2004). Epidemiology and pathophysiology of premenstrual syndrome. Obstetrics and Gynecology Research, 21(2), 112–120. doi: 10.1080/0951359060116796901

Fahoum, M., & Ross, K. (2023). An Integrative Approach for improving and Managing Premenstrual Syndrome (PMS) and Premenstrual Dysphoric Disorder (PMDD): a case report. Current Research in Complementary & Alternative Medicine, 7(4). https://doi.org/10.29011/2577-2201.100211

Farasati, N., Siassi, F., Koohdani, F., Qorbani, M., Abashzadeh, K., & Sotoudeh, G. (2015). Western dietary pattern is related to premenstrual syndrome: a case–control study. British Journal of Nutrition, 114(12), 2016-2021.

Farideh Bastani and Sara Hashemi (2012) Effects of a web based lifestyle education on general health and severity of the symptoms of Premenstrual Syndrome (PMS) among female students: a randomized controlled trial Procedia - Social and Behavioral Sciences, 46 3356 – 3362.https://doi.org/10.1017/S0007114515003943

Foley, A., Burgell, R., Barrett, J. S., & Gibson, P. R. (2014). Management strategies for abdominal bloating and distension. Gastroenterology & Hepatology, 10(9), 561–571.

Friedman, S. H., Tamburello, A. C., Kaempf, A., & Hall, R. C. (2019). Prescribing for women in corrections. J Am Acad Psychiatry Law, 47(4), 476-485.https://doi.org/10.29158/JAAPL.003885-19

Garg, P. (2014). Correlation of symptoms of premenstrual syndrome in Indian Set-Up. American Journal of Health Research, 2(3), 102. https://doi.org/10.11648/j.ajhr.20140203.15

Gatchel, R. J., Peng, Y. B., Peters, M. L., Fuchs, P. N., & Turk, D. C. (2007). The biopsychosocial approach to chronic pain: scientific advances and future directions. Psychological Bulletin, 133(4), 581-624. https://doi.org/10.1037/0033-2909.133.4.581

Geneen, L. J., Moore, R. A., Clarke, C., Martin, D., Colvin, L. A., & Smith, B. H. (2017). Physical activity for the management of chronic pain in adults: A systematic review and meta-analysis of randomized controlled trials. Pain, 158(11), 2119–2130. https://doi.org/10.1097/j.pain.0000000000001024

Ghanbari, Z., Manshavi, F., & Dashtdar, D. (2008). Effects of structured exercise programs on premenstrual syndrome symptoms in adolescents. Journal of Reproduction and Infertility, 12(4), 225–230. https://jfrh.tums.ac.ir/index.php/jfrh/article/view/53/53

Gordon-Smith, K., Perry, A., Di Florio, A., Craddock, N., Jones, I., & Jones, L. (2024). Associations between lifetime reproductive events among postmenopausal women with bipolar disorder. Archives of Women S Mental Health. https://doi.org/10.1007/s00737-024-01533-2

Groves, M. A., Moore, R. A., Colvin, L. A., & Smith, B. H. (2018). Anxiety in chronic pain populations: A comprehensive review. Journal of Pain Research, 11, 239–248. https://doi.org/10.2147/JPR.S169648

Gupta, R., Lahan, V., & Bansal, S. (2012). Subjective sleep problems in young women suffering from premenstrual dysphoric disorder. North American journal of medical sciences, 4(11), 593–595. https://doi.org/10.4103/1947-2714.103326

Hagen, K. B.,(2012). The effects of multidisciplinary rehabilitation for chronic pain patients: A systematic review and meta-analysis of randomized controlled trials. Pain, 153(10), 2011-2021. https://doi.org/10.1016/j.pain.2012.05.030

Halbreich U. (2007) Clinical diagnostic criteria for premenstrual syndrome and guidelines for their quantification for research studies: Gynecol endocrinol

Haleem, D. J., & Chakrabarti, A. (2010). The role of magnesium in premenstrual syndrome. European Journal of Clinical Nutrition, 64(8), 1044-1050.

Harel Z. Dysmenorrhea in adolescents and young adults: etiology and management. J Pediatr Adolesc Gynecol. 2006 Dec;19(6):363-71. doi: 10.1016/j.jpag.2006.09.001. PMID: 17174824.

Hoffman, M. A.,(2010). The role of rehabilitation in improving quality of life for patients with chronic pain. Journal of Rehabilitation Research and Development, 47(1), 33-42. https://doi.org/10.1682/JRRD.2009.02.0036

Hofmann, S. G. (2012). The efficacy of cognitive behavioral therapy for anxiety disorders: A review of meta-analyses. Cognitive Therapy and Research, 36(5), 426–440. https://doi.org/10.1007/s10608-012-9476-x

Huang, Y., Zhou, R., Wu, M., Wang, Q., & Zhao, Y. (2014). Premenstrual syndrome is associated with blunted cortisol reactivity to the TSST. Stress, 18(2), 160–168. https://doi.org/10.3109/10253890.2014.999234

Iacovides, S., Avidon, I., & Baker, F. C. (2015). Does pain vary across the menstrual cycle? A review. European Journal of Pain, 19(10), 1389-1405. **https://doi.org/10.1002/ejp.714**

Jami, S. (2016). Nutritional factors influencing severity of premenstrual syndrome. International Journal of Nutrition, 9(5), 301–309.

Jayakody, K., Gunadasa, S., & Hosker, C. (2014). Exercise for anxiety disorders: systematic review. British journal of sports medicine, 48(3), 187–196. https://doi.org/10.1136/bjsports-2012-091287

Jensen, M. P., Johnson, M. E., Turner, J. A., Romano, J. M., Gertz, K. J., & Scherrer, P. (2015). Pain disability and its correlation with clinical and demographic variables in chronic pain patients. Journal of Pain, 16(5), 455–462. https://doi.org/10.1016/j.jpain.2015.01.006

Jensen, M. P., Romano, J. M., Turner, J. A., Good, A. M., Gertz, K. J., & Scherrer, P. (2014). Pain management interventions and their impact on pain disability and quality of life: A review of outcomes. Pain Medicine, 15(6), 1023–1032. https://doi.org/10.1111/pme.12456

Jensen, M. P., Turner, J. A., Romano, J. M., & Karoly, P. (2003). Cognitive-behavioral therapy and rehabilitation for chronic pain: A meta-analysis. The Journal of Pain, 4(5), 242–252. https://doi.org/10.1016/S1526-5900(03)00333-5

Kaur, J., Kaur, K., Kaur, J. K., Kaur, K., Singh, L., Kaur, L., Kaur, M., Kaur, A. S. M., Kaur, B. M., Kaur, H. M., Kaur, R. S. M., & Jacob, T. (2016). Effectiveness of Planned Teaching Programme on Knowledge Regarding Life Style Changes in Prevention of Pre-Menstrual Syndrome among Adolescent Girls in A Selected School at Dehlon, Ludhiana. International Journal of Nursing Education and Research, 4(3), 327. https://doi.org/10.5958/2454-2660.2016.00058.2

Khodakarami, B., Babakhani, N., Masoumi, S. Z., Mohagheghi, H., & Farhadian, M. (2023). The Effect of Cognitive-behavioral Counseling on the Resilience of Female Adolescents with Premenstrual Syndrome: A Randomized Controlled Trial. PubMed, 11(4), 237–246. https://doi.org/10.30476/ijcbnm.2023.99040.2276

Khoury, B., Author, B. B., Author, C. C., & Author, D. D. (2013). Mindfulness-based interventions for anxiety and depression: A systematic review and meta-analysis. Journal of Psychosomatic Research, 74(6), 535–544. https://doi.org/10.1016/j.jpsychores.2013.03.003

Kim, S., Park, H., Lee, H., & Lee, H. (2011). Acupuncture for premenstrual syndrome: a systematic review and meta-analysis of randomised controlled trials. BJOG an International Journal of Obstetrics & Gynaecology, 118(8), 899–915. https://doi.org/10.1111/j.1471-0528.2011.02994.x

Kimberly Ann Yonkers. (2008) Premenstrual syndrome: Lancet. 371(9619): 1200–1210. doi:10.1016/S0140-6736(08)60527-9. 2017 Feb;4(2):146-158.

Kroenke, K., Wu, J., Bair, M. J., Krebs, E. E., Damush, T. M., & Tu, W. (2010). The association of depression and pain with disability in primary care: A systematic review. Journal of Clinical Psychiatry, 71(5), 722–729. https://doi.org/10.4088/JCP.09m05332blu

Kuehner, C. (2017). Why is depression more common among women than among men? The Lancet Psychiatry, 4(2), 146–158. https://doi.org/10.1016/S2215-0366(16)30263-2

Kues, J. N., Janda, C., Kleinstäuber, M., & Weise, C. (2015). How to measure the impact of premenstrual symptoms? Development and validation of the German PMS-Impact Questionnaire. Women & Health, 55(9), 1121–1139. https://doi.org/10.1080/03630242.2015.1118734

Kumar, A., Dhadwal, Y., Yadav, V., & Sharma, B. (2024). A cross-sectional study of knowledge, taboos, and attitudes towards menstruation. Ethnicity

& Health, 29(2), 208–219. https://doi.org/10.1080/13557858.2023.2293450

Lathavenkatesan, M., & Annapoorani, B. (2019). Comparative analysis of exercise and dietary interventions on Premenstrual Syndrome. Women's Health Journal, 15(3), 125–133. https://doi.org/10.1080/13697137.2019.1673891

Lee, L. K., Chen, P. C. Y., Lee, K. K., & Kaur, J. (2006). Menstruation among adolescent girls in Malaysia: a cross-sectional school survey. Singapore Medical Journal, 47(10), 869–874. https://pubmed.ncbi.nlm.nih.gov/16990962/

Liu X, Li R, Wang S, Zhang J. Global, regional, and national burden of premenstrual syndrome, 1990-2019: an analysis based on the Global Burden of Disease Study 2019. Hum Reprod. 2024 Jun 3;39(6):1303-1315. doi: 10.1093/humrep/deae081. PMID: 38689567.

MacGregor, E. A. (2008). Estrogen and Migraine: Correlations and Prevention. Headache: The Journal of Head and Face Pain, 48(s2), S99–S107. https://doi.org/10.1111/j.1526-4610.2008.01202.x

Maddineshat, M., Tabatabaeichehr, M., Ghanizadeh, Abotorabi-Zarchi, & Mohammadi. (2016). Effectiveness of Group Cognitive-Behavioral therapy on symptoms of Premenstrual Syndrome (PMS) [Original Article]. Iranian Journal of Psychiatry, 30–36. https://www.researchgate.net/publication/302414588

Maged, A. M., Abbassy, A. H., Sakr, H. R. S., Elsawah, H., Wagih, H., Ogila, A. I., & Kotb, A. (2018). Effect of swimming exercise on premenstrual syndrome. Archives of Gynecology and Obstetrics, 297(4), 951–959. https://doi.org/10.1007/s00404-018-4664-1

Maleki, F., Pourshahbaz, A., PhD, Asadi, A., & Yoosefi, A. (2014). The impact of Premenstrual Disorders on Health-related Quality of Life (HRQOL). University of Social Welfare and Rehabilitation Sciences.

Mangione, C. M.,(2007). Improvement in physical functioning in patients with chronic pain: Effects of rehabilitation interventions. Journal of Clinical Rehabilitation, 21(7), 483-492. https://doi.org/10.1177/0269215507073457

Manzoli, L., Di Natale, A., & Congedo, G. (2012). Effects of yoga on premenstrual syndrome: A systematic review and meta-analysis. BMC Complementary and Alternative Medicine, 12, 49. https://doi.org/10.1186/1472-6882-12-49.

Maurya, P., Meher, T., & Muhammad, T. (2022). Relationship between depressive symptoms and self-reported menstrual irregularities during adolescence: evidence from UDAYA, 2016. BMC Public Health, 22(1), 758.

Mazaheri Asadi, S.,(2020). Premenstrual Syndrome and its impact on school absenteeism: A study in Bangladesh. Adolescent Health Review, 8(2), 95–102.

McCracken, L. M., Vowles, K. E., & Eccleston, C. (2005). The role of cognitive and emotional factors in the treatment of chronic pain. Journal of Behavioral Medicine, 28(2), 137–147. https://doi.org/10.1007/s10865-005-3227-2

Mehdi Ranjbaran, Reza Omani Samani., Amir Almasi-Hashiani, Pegah Matourypour, Ashraf Moini: Prevalence of premenstrual syndrome in Iran International Journal of Reproductive BioMedicine Vol. 15. No. 11. pp: 679-686, November 2017

Meir Steiner, Miki Peer, Eva Palova, Ellen W. Freeman, Mary MacDougall, Claudio N. Soares, The premenstrual symptoms screening tool revised for adolescent of severe PMS and premenstrual dysphoric disorder in adolescent, Archives of Women's Mental Health. 2011; 14(1): 77–81

Miller, D. R., (2016). Energy and fatigue in chronic pain populations: Regional and cultural variations. The Journal of Pain, 17(5), 502-510. https://doi.org/10.1016/j.jpain.2015.12.001

Miller, K. B., & Grant, W. G. (2004). Guidelines for diagnosis and managing Premenstrual Syndrome. American Journal of Obstetrics and Gynecology, 19(7), 456–464.

Moseley, G. L. (2003). A review of the role of neuroplasticity in chronic pain: Implications for treatment. Journal of Pain, 4(6), 303-311. https://doi.org/10.1016/S1526-5900(03)00347-3

Nagi P., Mishra A., Lakhera P. (2018). Menstrual abnormalities and their association with lifestyle pattern in adolescent girls of Garhwal, India. Journal Family Medicine and Primary Care**7**:804-08.

Nandakumar, H., Kuppusamy, M., Sekhar, L., & Ramaswamy, P. (2023). Prevalence of premenstrual syndrome among students–Stress a potential risk factor. Clinical Epidemiology and Global Health, 23, 101368.

Oboza, P., Ogarek, N., Wójtowicz, M., Rhaiem, T. B., Olszanecka-Glinianowicz, M., & Kocełak, P. (2024). Relationships between premenstrual syndrome (PMS) and diet composition, dietary patterns and eating behaviors. Nutrients, 16(12), 1911.

Omidvar, S., Bakouei, F., Amiri, F. N., & Begum, K. (2016). Primary Dysmenorrhea and Menstrual Symptoms in Indian Female Students: Prevalence, Impact and Management. Global journal of health science, 8(8), 53632. https://doi.org/10.5539/gjhs.v8n8p135

O'Neill, M., Hoh, K., & Pasternak, D. (2009). The effects of cognitive-behavioral therapy on premenstrual syndrome: A review of the literature. Behavioral Therapy, 40(4), 1056-1064.

Parikh, M. N., Parikh, N. C., & Parikh, S. K. (2015). Premenstrual Dysphoric Disorder in adolescent girls in Western India. GMJ, 70(1), 65–69. https://doi.org/10.18203/2320-6012.gmj20153096

Pasquale, A., Uršulin–Trstenjak, N., & Matošević, I. (2022). The Role of Nutrition in Achieving Hormonal Balance in Women. Horizons-International Scientific Journal, 31(2), 365-376. https://doi.org/10.20544/HORIZONS.A.31.2.22.P28

Patel, K. R. (2018). Physical activity and psychological interventions in chronic pain: Improvements in health perceptions. Pain Management, 8(3), 255–264. https://doi.org/10.2217/pmt-2017-0033

Patel, K. R., (2015). Depression and anxiety in chronic pain populations: Baseline characteristics and psychological implications. Journal of Pain Research, 8, 213-220. https://doi.org/10.2147/JPR.S84922

Ramya S, Rupavani K, Bupathy A. Effect of educational program on premenstrual syndrome in adolescent school girls. Int J Reprod Contracept Obstet Gynecol 2014;3:168-71.

Rapkin, A. J., & Winer, S. A. (2009). Premenstrual syndrome and premenstrual dysphoric disorder: quality of life and burden of illness. Expert Review of Pharmacoeconomics & Outcomes Research, 9(2), 157–170. https://doi.org/10.1586/erp.09.14

Ravichandran, S., and Janakiraman, R. (2018). Aerobic exercises as an intervention for Premenstrual Syndrome in adolescents. Journal of Exercise and Fitness, 20(1), 56–62.

Ravichandran, S., and Janakiraman, R. (2019). Long-term benefits of aerobic exercise on menstrual health. International Journal of Sports Medicine, 34(6), 312–319.

Rethorst, C. D., and Trivedi, M. H. (2013). Evidence-based recommendations for the prescription of exercise for major depressive disorder. Journal of Psychiatric Practice, 19(3), 204–212.

Roca, C. A., Schmidt, P. J., Altemus, M., Deuster, P., Danaceau, M. A., Putnam, K., & Rubinow, D. R. (2003). Differential Menstrual Cycle

Regulation of Hypothalamic-Pituitary-Adrenal Axis in Women with Premenstrual Syndrome and Controls. The Journal of Clinical Endocrinology & Metabolism, 88(7), 3057–3063. https://doi.org/10.1210/jc.2002-021570

Ruprai, R., Seth, T., Singh, R., Tandon, A., & Sethi, R. (2015). Nutritional status and menstrual health in South Delhi. Journal of Public Health Research, 11(5), 176–183.

Sahoo, J., Singh, S. V., Gupta, V. K., Garg, S., & Kishore, J. (2015). Do socio-demographic factors still predict the choice of place of delivery: A cross-sectional study in rural North India. Journal of Epidemiology and Global Health, 5(S1), S27. https://doi.org/10.1016/j.jegh.2015.05.002

Sarkar AP, Mandal R, Ghorai S. Premenstrual syndrome among adolescent girl students in a rural school of West Bengal, India. Int J Med Sci Public Health 2016;5:773-6.

Shamnani, G., Gupta, V., Jiwane, R., Singh, S., Tiwari, S., & Bhartiy, S. (2018). Prevalence of premenstrual syndrome and premenstrual dysphoric disorder among medical students and its impact on their academic and social performance. National Journal of Physiology Pharmacy and Pharmacology, 8(9), 1205. https://doi.org/10.5455/njppp.2018.8.0415728042018

Sharma, P., Malhotra, C., Taneja, D. K., & Saha, R. (2008). Problems related to menstruation amongst adolescent girls. Indian journal of pediatrics, 75(2), 125–129. https://doi.org/10.1007/s12098-008-0018-5

Singh A, Kiran D, Singh H, Nel B, Singh P, Tiwari P. Prevalence and severity of dysmenorrhea: A problem related to menstruation, among first and second year female medical students. Indian J Physiol Pharmacol 2008;52:389-97.

Smeets, R. J. E. M., Vlaeyen, J. W. S., Hidding, A., Kester, A. D. M., van der Heijden, G. J. M. G., van Geel, A. C. M., & Knottnerus, J. A. (2006). The efficacy of physical and psychological interventions for chronic pain patients: A systematic review. Pain, 124(3), 223–229. https://doi.org/10.1016/j.pain.2006.06.017

Smith, A. J.,(2018). Variability in pain-related disability in chronic pain populations. The Journal of Pain, 19(3), 275-284. https://doi.org/10.1016/j.jpain.2017.09.001

Spielberger, C. D., et al. (2014). Anxiety and stress in chronic pain patients: An examination using the Beck Anxiety Inventory. Journal of Clinical Psychology, 70(3), 304-312. https://doi.org/10.1002/jclp.22068

Stanley, M. A.,(2003). Cognitive-behavioral treatment for anxiety disorders in individuals with chronic pain: A review of the evidence. Journal of Pain, 4(7), 397-407. https://doi.org/10.1016/S1526-5900 (03)00382-8

Steer, R. A., Rissmiller, D. J., & Beck, A. T. (2000). Use of the Beck Depression Inventory-II with depressed geriatric inpatients. Behaviour research and therapy, 38(3), 311–318. https://doi.org/10.1016/s0005-7967(99)00068-6

Sujata Deo1, Pratibha Kumari, Swati Vishwakarma, Pooja Mahour (2024) :Associated Factors of Premenstrual Syndrome (PMS) and Premenstrual Dysphoric Disorder (PMDD) in Adolescent and Young Women in North Indian PopulationSSR Institute of International Journal of Life Sciences ISSN (O): 2581-8740 | ISSN (P): 2581-8732 DOI: 10.21276/SSR-IIJLS.2024.10.5.15

Sunitha, S. (2011). Hormonal influences on emotional well-being during adolescence. Endocrinology and Adolescent Health, 23(4), 410–420.

Tabassum, S., Afridi, B., Aman, Z., Tabassum, W., & Durrani, R. (2005). Premenstrual syndrome: frequency and severity in young college girls. JPMA. The Journal of the Pakistan Medical Association, 55(12), 546–549.

Taghizadeh Z., Shirmohammadi M., Mohammad A., Abbas M. (2008). Effect Of Premenstrual Syndrome on Quality of Life In Adolescent Girls. Iran Journal Psychiatry **3**:105-109.

Tanaka, H., (2019). Cultural influences on fatigue perception and management in chronic pain patients. Pain Research and Management, 2019, 6340321. https://doi.org/10.1155/2019/6340321

Tarannum F, Khalique N, Eram U. Premenstrual syndrome: Prevalence, symptoms, and associated risk factors among adolescent girls in Aligarh, Uttar Pradesh. Indian J Public Health 2021;65:396-9

Taylor, D. (2006). Cognitive behavioral therapy for Premenstrual Syndrome: A clinical perspective. Psychology and Health, 18(3), 98–106.

Thakre SB, Thakre SS, Ughade S, Thakre AD. Urban-rural differences in menstrual problems and practices of girl students in Nagpur, India. Indian Pediatr 2012;49:733-6.

Treloar SA, Heath AC, MARTIN NG. (2022) Genetic and environmental influences on premenstrual symptoms in an Australian twin sample. Psychological Medicine. 2002;32(1):25-38. doi:10.1017/S0033291701004901

Turk, D. C.,(2011). Pain and disability management in chronic pain conditions: Targeted interventions and outcomes. Clinical Journal of Pain, 27(4), 299-307. https://doi.org/10.1097/AJP.0b013e318205659f

U.S. Department of Veterans Affairs. (2005). The management of chronic pain in veterans: A comprehensive treatment approach. Journal of Rehabilitation Research and Development, 42(2), 97-104. https://doi.org/10.1682/JRRD.2005.03.0058.

Upadhyay, M., Mahishale, A., & Kari, A. (2023). Prevalence of premenstrual syndrome in college going girls-A cross sectional study. Clinical Epidemiology and Global Health, 20, 101234.

Ussher, J. M. (2008). Challenging the positioning of Premenstrual change as PMS: The Impact of a Psychological Intervention on Women's Self-Policing. Qualitative Research in Psychology, 5(1), 33–44. https://doi.org/10.1080/14780880701863567

Van Eijk, A. M., Sivakami, M., Thakkar, M. B., Bauman, A., Laserson, K. F., Coates, S., & Phillips-Howard, P. A. (2016). Menstrual hygiene management among adolescent girls in India: a systematic review and meta-analysis. BMJ Open, 6(3), 1–12. doi:10.1136/bmjopen-2015-010290

Van Eijk, A. M., Sivakami, M., Thakkar, M. B., Bauman, A., Laserson, K. F., Coates, S., & Phillips-Howard, P. A. (2016). Menstrual hygiene management among adolescent girls in India: a systematic review and meta-analysis. BMJ open, 6(3), e010290. https://doi.org/10.1136/bmjopen-2015-010290

Ware, J. E.,(2000). SF-36 health survey update: Methodological developments and refinements. Journal of Clinical Epidemiology, 53(7), 431-440. https://doi.org/10.1016/S0895-4356(00)00273-4

Watkins, R. S.,(2019). Psychiatric comorbidities in chronic pain patients: Impact of severe pain on depression and anxiety scores. Pain Medicine, 20(7), 1348-1355. https://doi.org/10.1093/pm/pnz080

Williams, A. C. D. C.,(2012). Cognitive-behavioral therapy for chronic pain in adults: A meta-analysis of randomized controlled trials. Pain, 153(3), 712-723. https://doi.org/10.1016/j.pain.2011.12.012

Wise, C.,(2019). Group Cognitive Behavioural Therapy for Premenstrual Syndrome: Outcomes and effectiveness. Cognitive Behavioral Therapy Journal, 9(2), 67–79.

World Health Organization. (1992). Definition and classification of premenstrual syndrome. International Classification of Diseases Manual, 10th Edition, 145–150.

Yonkers, K. A., & Simoni, M. K. (2018). Premenstrual disorders. American journal of obstetrics and gynecology, 218(1), 68-74.

Zhao, F., Fu, Q., Kennedy, G. A., Conduit, R., Zhang, W., & Zheng, Z. (2021). Acupuncture as an Independent or Adjuvant Management to Standard Care for Perimenopausal Depression: A Systematic Review and Meta-Analysis. Frontiers in Psychiatry, 12. https://doi.org/10.3389/fpsyt.2021.666988

Zoghi, M., Ghaffari, M., & Salari, N. (2010). The effect of aerobic exercise on premenstrual syndrome symptoms in women: A systematic review and meta-analysis. Journal of Sport Science and Medicine, 9(3), 336-345.

www.ingramcontent.com/pod-product-compliance
Lightning Source LLC
LaVergne TN
LVHW082349200926
848481LV00012B/3502